SAN DIEGO

2.8 ANGSTROMS

IS IT A QUIRK OF NATURE TO CONFUSE MAN
OR DOES MAN UNWITTINGLY CONFUSE NATURE

2.8 ANGSTROMS

The Unifying Force of G and c

Kenneth G. Salem

Salem Books
Johnstown, PA

2.8 ANGSTROMS, First Edition

Copyright © 1990 by Kenneth G. Salem,
All rights reserved. Printed in the
United States of America.
Salem Books, 1990

Salem, Kenneth G
 2.8 Angstroms

ISBN 0-9625398-0-5

Contents

	Preface	vii
1.	Introduction	1
2.	The Philocity of Light	7
3.	The Not So Constant Constancy of Light's Velocity	13
4.	The Three Truths	19
5.	"Why" The Velocity of Light Is NOT Constant	33
6.	"Why" $E=mc^2$	39
7.	"Why" The Universe Expands	43
8.	The Universe In Numbers	49
9.	On The Way Light Propagates	63
10.	How Mass Is Created	83
11.	What Holds It All Together	95
12.	Speculations	99
13.	GUT's and TOE's	107

Appendix I	Equations Utilizing a_u	111
Appendix II	Constants and Values	113
Appendix III	Symbols	115
Appendix IV	Mass-Energy Conversion Table	117
Bibliography		119

Preface

When I first realized the phenomenal speed with which light traveled, it fascinated me to no end. As time went on and I thought more and more about it, I consistently ended up with the same question—How can it be? How can it be, I thought, that anything in nature can just "up and go that fast," and yes, even a light beam's massless photons.

I was not denying that a light beam could, for example, propagate from Earth to Moon in one and a quarter seconds. There was no question in my mind that light and other forms of electromagnetic phenomena do indeed get from place to place at the enormous speed of 186,282 miles per second. But what the basic mechanism is that is responsible for this phenomenal feature of light was indeed itself the big question.

Those times when I thought to myself, "It Cannot Be", what I mean to imply is that it simply cannot be that anything in nature, *not even a light beam,* can develop such a phenomenal speed out of practically "sheer nothingness." However, this was not necessarily the specific feature about light which kept me persisting until I had to write this book.

What actually did hook me on the way in which light traveled, was the very unique and strange quality it possesses, in that *it always starts out from its source at 186,282 miles per second.* What this means, very simply, is that when a light beam leaves its source, say from a flashlight for example, the beam "never starts out from zero-velocity" to get up to speed of 186,282 miles per second, but rather, it leaves the *supposedly stationary flashlight* with an initial starting speed of 186,282 miles per second!

When it became sufficiently clear to me as to what this was actually saying about a very important characteristic of light, I decided that this was one thing I could not, *nor would not accept.* I simply decided that nothing in nature (not even a quanta of light) *could go on a journey without having to start out.*

After I became well apprised of this particular feature of light's qualities, I then made it part of my life's work to search for the reason "Why" light propagated in such a strange manner. And now, after more than forty years of delving into the subject and contemplating light's mechanism, I decided to write about it. Why? Because I believe this odd characteristic of light has a "forthright and common sense explanation to it."

Two-Point-Eight Angstroms not only attempts to explain "Why" light always leaves its source without having to accelerate to get to speed, Two-Point-Eight Angstroms also attempts to explain the very basis of *Gravity* itself. It will show where gravity actually originates from, and brings together *Electromagnetic Interactions and Gravity* to show the relationship between the two as being due to one and the same Basic Mechanism.

In addition to explaining the "Single Mechanism" responsible for both Gravity and Electromagnetism, 2.8 Angstroms will also show that this same mechanism can serve to explain:

1. Why $E=mc^2$
2. Why the Hubble Law and the Hubble Constant are what they are
3. Why it has been virtually impossible for scientists to determine (with precision) the Position and Momentum of the electron simultaneously
4. Why the Wave-Particle Duality exists
5. What Creates Mass
6. Why Plank's Constant is precisely what it is
7. How Quasars come about
8. The 3° Cosmic Background Radiation—possibly another explanation for it
9. And, amongst other things, 2.8 Angstroms will attempt to show why Michelson and Morley—in their famous experiments of the 1880's—were unable to confirm the "Principle of the Addition of Velocities" as regards the Velocity of Light . . .

There is however, one additional reason for writing this book. In February 1989 I had mailed an "Eight-page paper" to more than 10,000 physicists throughout the United States—especially those working in Particle Physics, Nuclear Physics, and Astrophysics. The paper described the theory which this book purports to describe. However, in that "paper", the *physical units* of the most important equations did not come out correctly. As a result, I received many letters from physicists pointing this matter out to me. Consequently, I ran a one-third page "Ad" in the May 89 issue of "Physics Today" in which I corrected the most significant of the dozen or so equations originally shown.

I was also rebuffed by some (and rightly so) for stating that "Evidence Exists" to support the theory I had expounded upon, but in fact, did not describe *what that evidence was.* And so, I felt even more compelled to write this book in order that this important omission may also be remedied.

1

Introduction

Einstein's Theories of Relativity set physics reeling on a whole new course, one which radically changed the scientific community's concept of Space and Time. His contributions to science included, amongst other things, his famous equation on Mass and Energy, $E=mc^2$ which is still today considered the most famous equation of all time.

I believe however, that in spite of the many good and positive works Einstein contributed, he may have "unwittingly" steered the scientific community on an erroneous course when he made the far reaching assumption that *The Speed of Light Is An Absolute And Unchanging Constant Of Nature Throughout All Space and Time*. He later raised his assumption to a *postulate,* and thus, it became a major turning point in the way we think about physics.

In this book it will be advocated quite to the contrary, that The Speed Of Light Is *NOT* An Absolute And Unchanging Constant Of Nature, but that there appears to exist an "Outside Force" throughout all Space and Time which *does in fact cause light's velocity to continuously increase with time*. This Outside Force apparently creates a Universal Constant of Acceleration, a_u, of approximately 2.8×10^{-10} meter (2.8 Angstroms) per second per second which is imposed upon all material bodies in the Universe (from electrons and protons to planets and stars).

Consequently, the *ensuing and varied velocities* of all material bodies throughout the Universe become the vehicle, the means in which Light and all other forms of Electromagnetic Phenomena derive their energy of propagation. In reality then, it is the material bodies of the Universe (the protons, neutrons, and electrons, etc. comprising the galaxies of stars and planets) which actually possess the *Real and Basic velocity*—the Energy which the Velocity of Light itself derives from—increasing as they do, by a mere 2.8 Angstroms for each second of time. So that in 34 Billion Years, for example, the materials in our own Galaxy have accelerated at this constant rate to their present velocity of 186,282 miles per second, while having traversed a distance of Seventeen-Billion of our present light years in *one general direction from time-zero* ($t=o$), where it is generally believed the Center of Mass of our Universe originated.

Therefore, for any Isolated Spherical Body of Uniform Density, p, and Radius, r, it will be shown that this Universal Constant of Acceleration, a_u is always equal to (F/m_r), where (F) is the magnitude of g in Newtons per kilogram at a body's surface, and (m_r) represents a body's (MASS of ONE SQUARE METER RADIUS), a *NEW UNIT* which reflects a body's RADIUS in *CUBIC METERS* rather than its customary radius in (METERS). So that, "even for a proton," which has a radius of about 10^{-15} meter, its (Square Meter Radius volume) would be 10^{-15} Cubic Meter (instead of its actual 10^{-45} m^3) for purposes of determining both, a_u, and the magnitude of g at its surface, F. For example, in using the physical parameters of the proton, we determine for a_u

$$a_u = F_p/m_r \tag{1}$$
$$a_u = 1.2 \times 10^{-7} \text{N} / (3.1 \times 10^{17} \text{ kg/m}^3)(1.4 \times 10^{-15} \text{ m}^3)$$
$$a_u = 1.2 \times 10^{-7} \text{N} / 4.3 \times 10^2 \text{ kg} = 2.791 \times 10^{-10} \text{m/s}^2$$

This equation ($a_u = F/m_r$) maintains a constant ratio between a body's MASS OF ONE SQUARE METER RADIUS and the MAGNITUDE of g at its surface. This ratio (2.791×10^{-10} m/s^2) is the UNIVERSAL CONSTANT of ACCELERATION, a_u, which actually creates that body's Gravitational Forces.

It follows then, that the magnitude of g at any body's surface would be ($F = m_r a_u$); and for a proton, $F_p = 1.2 \times 10^{-7}$ N.

The constant (a_u) is calculated to be $2.79062594 \times 10^{-10}$ meter (otherwise 2.791 Angstroms) per second per second. However, for all practical purposes a_u will usually be referred to as 2.8 Angstroms throughout the book.

I will also attempt to show that the "Hubble Law" has a direct relation to the Universal Constant of Acceleration of 2.8 Angstroms per second per second. And I will show also that there is a good "common sense" explanation why the Michelson-Morley experiments of 1887 did not result in verifying the variations to the velocity of light which they so expected to find. The sketches relating to this are designed to help overcome a bias which has existed about the manner in which light propagates.

In the latter part of the book I will *highly speculate,* amongst other things, about the real nature of *Quasars,* and the *Three Degree Cosmic Background Radiation.* The theories on these and others will, I hope, be interesting. However, I say (speculate) only because the heart of the book is based upon Three Hard and Proven Facts of the physical sciences—*Three Truths*—as I will refer to them from time to time throughout the book. These are "indeed" three of the most important Cosmological Discoveries ever made, and should prove to be very supportive in lending credence to what will be advocated in this book, most specifically: *That it is we ourselves that are physically moving through space at the velocity we attribute to light; and that Gravity itself is the direct result of our own constant acceleration through space.* That acceleration of 2.8×10^{-10} meter per second per second is such an infinitesimal amount (only one-ninety-millionth of an inch per second per second) that I fully realize it may be incomprehensible for one to envision that such enormous and massive bodies as stars like the Sun, and planets like the Earth could actually be accelerating this very minuscule distance (equal to the diameter of an average size molecule) for each second of time as they fly off through space at their present velocity of 186,282 miles per second. From all the facts available however, this seems to be just the picture.

Universal Gravity, G, may well then be attributable to all bodies "Inherently Accelerating" at a constant rate of 2.8 Angstroms per second per second. And Electromagnetic phenomena would then be directly attributable to a body's present (basic) velocity through space. So that when we compare the Electromagnetic Interactions to the very minuscule Gravitational Interaction, *both of which appear to derive from the same basic mechanism,* we do indeed come up with a ratio, R, as so often has been shown to exist between the two—a ratio of 10^{36} to one. For when one divides the square of this Universal Constant of Acceleration, a_u^2, into the square of our present velocity of light, c^2, (*each for an elapsed time of one-second*) we find that the Electromagnetic Interaction is indeed a *trillion trillion trillion* times more powerful than the Gravitational Interaction, both being a direct consequence of the Universal Constant of Acceleration, such that

$$R = c^2/(a_u/1.0 \text{ Sec})^2 \qquad (2)$$
$$R = 8.988 \times 10^{16}/7.788 \times 10^{-20}$$
$$R = 1.154 \times 10^{36}$$

This then should provide the relationship much sought after between Electromagnetic Interactions and Gravitational Interactions, both being simply and directly related to the Universal Constant of Acceleration and our own present velocity through space. Consequently, this ratio, *which of course determines our age,* should grow larger with time since a_u remains constant while c constantly increases with time; so that at present, the approximate value of c would be determined by the "product" of the length of time, t, a body has been accelerating since t=0, (1.074 × 10^{18} seconds in our own case) and the body's rate of acceleration, a_u such that for the Milky Way, MW, we would have a present velocity for "c" of

$$c_{MW} = a_u t_{MW} \qquad (3)$$
$$c_{MW} = (2.791 \times 10^{-10} \text{m/s}^2)(1.074 \times 10^{18} \text{sec})$$
$$c_{MW} = 2.998 \times 10^8 \text{m/s}$$

and for our own time of 34-billion years here at the Milky Way position, our own real velocity, v_{MW}, would simply be determined by the product of our present time, t, in "seconds," and the Universal Constant of Acceleration, a_u, in exactly the same manner we determine "c" above for the Milky Way, so that

$$v_{MW} = a_u t_{MW} \qquad (4)$$
$$v_{MW} = (2.791 \times 10^{-10} \text{m/s}^2)(1.074 \times 10^{18} \text{sec})$$
$$v_{MW} = 2.998 \times 10^8 \text{m/s}$$

These then, naturally determine both, *the real velocity of Earth through space, and it's local velocity of light at this particular epoch of space-time* in relation to the beginning of the Universe's expansion at t=0.

2

The Philocity of Light

How many mysteries of the Universe might be resolved if it were realized that the particles comprising the Earth (electrons, protons, neutrons, etc.) have been accelerating through space in *one general direction* at a constant rate of 2.8 Angstroms per second per second *for the past 34 billion years* and are presently up to a speed of 186,282 miles per second? That's right, the speed of light!

Indeed, if this is true, then both the propagation of light itself, and its precise speed would be direct consequences of the velocity of solid bodies through space. And as these solid bodies, such as our own Earth, for example, continue to accelerate to higher speeds, ever, ever so slowly, so too would the speed of light, *mimicking precisely the velocity of the Earth through space.*

And what of the mystery of Gravity? Would it not finally show itself for what it really is? For if in fact the elementary particles which make up the Earth, the Galaxies, etc., are constantly accelerating through space at 2.8 Angstroms per second per second, then that *"acceleration itself"* must surely be responsible for the *"gravitational attraction"* that holds us to the Earth like a magnet.

How can we find justification for all of this? One key to it is wrapped up in the knowledge which we presently possess of an Expanding Universe, for there is no better tell-tale sign to indicate that the Earth and all other material bodies in the Universe are constantly accelerating through space, than the direct evidence—first discovered by Edwin Hubble in 1929—that the Universe is in a state of constant, and "apparently," Uniform Expansion.

Hubble's value for this rate of expansion was originally estimated to be about 550-kilometers per second per megaparsec. This is about 170 kilometers (100-miles) per second per million light years. However, its present revised value is now only about 10% of his original estimate and is believed to be somewhere nearer to 18-kilometers (11-miles) per second per million light years, (or 57 kilometers per second per megaparsec) of which the megaparsec, *3.2 million light years*, is the unit of distance on the cosmological scale preferred by astrophysicists.

Now if we were to work this rate of expansion of 18 km per second per million light years towards smaller and smaller increments of space, *instead of towards higher and higher increments as is usually the case,* then when we finally get as close in to only one (1) light-second (186,282 miles), the expansion rate would hypothetically work out to be approximately 5.6×10^{-10} meter (5.6 Angstroms) per second per light-second. However, since it is here advocated that the velocity of light "Is Not" after all constant, but instead *constantly increases with time*, then naturally, the light-year distance scale must also increase with time—so that the Hubble Constant of 18-km per second per million light-years should, in reality, only be one half its estimated value, or about 9-kilometers per second per million light-years, which of course reduces down to 2.8×10^{-10} meter (2.8 Angstroms) per second per light-second.

In any case, we will concentrate on the recessional velocity of Hubble's Constant for one (1) light-second, 186,282 miles in distance, (ignoring for the moment that recessional velocities are generally believed to apply only between clusters of galaxies) and here on in refer to Hubble's Constant as having a value of 2.8 Angstroms per second per light-second.

Now since I have contended that Gravity is strictly the result of a body's constant acceleration of 2.8 Angstroms per second per second through space, it must then be shown how this acceleration through space is actually related to the gravitational attraction which such bodies as the Earth and Moon experience.

Since it is already a proven fact that the effects of gravitation and acceleration are equivalent (thanks to Einstein's Principle of Equivalence), we can now consider what the force of attraction should be at or near the surface of the Earth if the Earth is indeed accelerating through space at 2.8 Angstroms (2.8×10^{-10} meter) per second per second.

So let us now consider, for example, the Earth with its *Square-Meter-Radius* of 6,371,315 *"cubic"* meters (a new unit as described in chapter one), and its density of 5,516 kilograms per cubic meter accelerating through space at a constant rate of 2.8×10^{-10} meter per second per second. If we then multiply the Earth's Volume of its Square-Meter-Radius, v_r by its density, p, we come out with a product for *Mass of the Radius*, m_r, of approximately 35 Billion Kilograms. We next multiply this (Square Meter Radius) product times the Universal Rate of Acceleration of 2.8×10^{-10} meter per second per second to arrive at the magnitude of g in Newtons per kilogram at the Earth's surface (F) of 9.8 Newtons.

My own corrected value for the Hubble Constant of 9 kilometers per second per million light years, which as stated, works out to about 2.8×10^{-10} meter per second per light second, would be the actual required value in order for the magnitude of g near the surface of the Earth to equal 9.8 Newtons per kilogram. And this constant, a_u, accounts of course for the g-force on or near the surface of all material bodies throughout the Universe, whether we are speaking of electrons and protons or planets and stars.

Here then, we have in effect the equivalence of what the "Principle of Equivalence" itself predicts: that is, that "One cannot distinguish between being at rest in a gravitational field, and being accelerated upward in a gravity free environment." This Principle of Equivalence was a very important part of Einstein's General Theory of Relativity, and has always been upheld, that is, there is no dispute whatsoever of this particular aspect of Einstein's General Theory.

What other mysteries of the Universe might also be resolved if indeed our galaxy, the Milkyway, is advancing through space at a speed heretofore strictly attributable to Light and other forms of Electromagnetic phenomena?

If Light does indeed derive its energy of motion from *the motion of the source* it is emitted from, this would then answer for certain the question as to *why nothing can ever exceed the speed of light*. The answer, naturally, would be evident. It is because Light simply *mimics* the velocity of anything that moves. Move Faster—and Light Moves Faster! Light, and other forms of Electromagnetic Interactions are energy carrying images, and images consist of photons, and therefore, they have no mass. Consequently, they actually shoot away (and or appear to shoot away by dropping off from solid bodies) with real velocities, and or apparent velocities strictly proportional to that of the body they are emitted from.

Additional mysteries which can be cleared up if indeed we are rushing off through space at the speed we attribute to light, would be in connection with Einstein's famous equation on Mass and Energy, $E=mc^2$. And of course, the big question: "Why does light propagate with one specific speed?" Answers to these two questions then become immediately apparent. The velocity of light, c, in the equation $E=mc^2$ is squared simply because this velocity is actually the *Real Velocity, v, of the Mass, m,* for which we are determining the rest energy, E.

As for why Light always appears to propagate at one specific speed, would that not be apparent if indeed it is we who are rushing off through space at the speed we attribute to light? In our own case here on the Earth, we and the Earth rush through space at a speed of 186,282 miles per second, although all along we are constantly picking up speed, accelerating at the minuscule rate of 2.8 Angstroms for each second of time. However, for all practical purposes, it would still be correct to state that light propagates with one specific speed—always at the speed with which we here on Earth and the Earth itself rush off through space—186,282 miles per second. But that is only for this particular moment of time. During the next moment, or second of time, our speed increases by about one ninety-millionth of an inch (2.8 Angstroms), and for each additional second of time our speed increases another 2.8 Angstroms, and so on and so on, ad infinitum. However, small as this acceleration rate is, wherein the Earth's velocity only amounts to an increase of about "one centimeter per year," it has accelerated each and every "body" in our galaxy to a velocity of 186,282 miles per second in the past 34-billion years (10^{18} seconds) at a constant rate of acceleration.

Technically speaking, there simply cannot be any such thing in nature then, *as a perfectly constant velocity!* For all practical purposes however, we can safely state that light always propagates with one specific speed, i.e. with one constant velocity.

3

The Not So Constant Constancy of Light's Velocity

What do we actually mean when we speak of the Constant Velocity of Light? Webster's definition of "constant" means *not to change*, i.e., *remaining the same.* The prime theme of this book is that the velocity of light is indeed "Not Constant," and that the only thing constant about it is its *constantly increasing velocity* by 2.8 Angstroms per second per second.

There are generally Three Modes in which the Constant Velocity of Light is viewed throughout the literature:

First, there is "constant" in the sense that light always propagates with one specific speed. In this sense of "constant," it never has to accelerate from zero velocity (i.e., from its presumed position of rest) to get up to its one speed of propagation of 186,282 miles per second. This evidently is what James Clerk Maxwell meant in teaching that the velocity of light was constant. He presumably had no knowledge of whether or not the velocity of light was an "absolutely unchanging" constant, for even with our present day technology it is very doubtful that we would be able to detect such a minuscule change in the velocity of light by only 2.8 Angstroms per second. This is only one ninety-millionth of an inch per second per second.

The main thing to be remembered when we speak of this Constancy of Lights Velocity is that this is indeed, *for all practical purposes* an accurate statement as such, and it should not be allowed to become entangled with other versions of the word "constant" in connection with "Light." Maxwell's meaning was and still is very correct. He stated in effect, that *Light Always Moves From Place To Place With One Speed!* Now if it does indeed increase its speed by only one part in one-thousand-quadrillion per second, he certainly had no idea of this, and he certainly wasn't implying anything of this nature when he said what he did about the way in which light propagates. He was simply referring to the fact that light's velocity (of the order of 3.0×10^8 meters per second) was the only velocity with which it moves about from place to place, and that it did not have to "accelerate" in order to achieve that speed. It was, in other words, *a Built In Speed*.

Another mode in which light is referred to as moving from place to place with a constant velocity is that which refers to the famous experiments of the 1880's by Michelson and Morley in which they set out to detect a variation of light's velocity due to the Earths 18-mile per second orbit around the Sun.

It is well known that Michelson and Morley failed to detect the expected variation in the velocity of light in accordance with "The Principle of the Addition of Velocities." Their experiments showed instead that no matter which direction they pointed their light beam in respect to the Earth's forward motion around the Sun, the velocity of light did not appear to show any variation whatsoever. As a result of their very meticulous experiments, it was then presumed that the velocity of light was *also constant in this other sense,* which, in effect, is quite different from that which Maxwell had first postulated.

So even in this sense Michelson and Morley's experiments showed (in yet another mode) that light always had the same speed no matter "which direction" the beam was aimed from a "moving" body. In other words, and according to Einstein, even if the moving body—the Earth in this example—were traveling around the Sun at 99% the speed of light, the results Michelson and Morley had gotten of the Earth's 18-mile per second orbit around the Sun would be the same result they would get if the Earth could indeed orbit the Sun at 99% the speed of light (184,419 miles per second instead of 18-miles per second). This type of reasoning provides a good example of what Einstein most likely meant when he denounced the use of *common sense* in contemplating the mechanism of the Universe.

What Michelson and Morley's experiments supposedly proved then, was not only did light move about from place to place at one specific speed as taught by Maxwell, but that the additional velocity supposedly imparted to the light beam from the Earth's moving body around the Sun did not make any difference whatsoever in the beam's rate of propagation. It neither gained in relative speed when aimed in the Earth's forward direction around the Sun, nor lost in relative speed when aimed at right angles to the Earth's direction around the Sun. The Principle of the Addition of Velocities *did not appear* to apply in the case of "light propagation."

So here we have a second version of what the Constant Velocity of Light actually means. I intend to show however that there is a good *common sense explanation* for the results which Michelson and Morley had gotten; and that the Principle of the Addition of Velocities *does in fact hold true with respect to light.*

Now for the third version of the Constant Velocity of

Light—*Einstein's Version*. This is the version, unfortunately, which I feel may have brought physics to somewhat of a standstill.

After Michelson and Morley's failure to detect a variation of the velocity of light, Einstein, professing not to have known of their work, supposedly played with the idea for some years as to why light's velocity did not adhere to the rules of the Principle of The Addition of Velocities. Finally, in the early part of 1900, Einstein made some far reaching assumptions about this failure of light's velocity to vary. What he said was that the reason light's velocity did not vary in those cases where it was expected to vary, that is, in accordance with the Principle of the Addition of Velocities, was because there was "No Ether" in the first place to carry the light. The *ether* was at that time believed to pervade all of space and was thought to be the medium for carrying the light waves—just as *air* is the medium for carrying sound waves. Secondly, and most importantly, Einstein said that *The Velocity of Light was an Absolute and Unchanging Constant Throughout All Space and Time*. He said, in fact, it was space and time which were suspect, which themselves were not constant; and that our concept of space and time simply was not accurate.

What Einstein was saying in effect, was that the speed of light never changes anytime, nor anywhere throughout the vast universe—not even by one-angstrom unit per year—not by any amount throughout all space and time!

There was no argument that light always propagates with only one speed; nor any argument "that it did not have to accelerate" to get to its sole propagating speed of 186,282 miles per second. This being so, and in addition to the fact that the Michelson-Morley experiments failed to detect any change in the speed of light when common sense dictated that they should, *Einstein finally reasoned that common sense is not always the answer*. He stated that past experiences tend to prejudice our thinking. This then, is when he decided that our concepts of Space and Time must not be totally accurate, henceforth, came his Special Theory of Relativity in 1905 in which he put a "Limiting Velocity" on the speed of light, and therefore, as I have stated previously, he may have *"unwittingly" steered the physics community on an erroneous course.*

The important factors then, which Einstein felt were absolutely necessary if he were going to be able to change the way we think about space and time, was that not only does the velocity of light always propagate at one specific speed *without having to accelerate* to get to that speed—but that the velocity of light was also an absolute and unchanging constant throughout all space and time. So that no matter how fast, nor in which direction an observer or a light source moved with respect to one another, the Principle of the Addition of Velocities was not, he said, not applicable to Light as it was with other phenomena of nature. This thinking, which I purposefully repeat once again, arose strictly as a consequence of his assumption (which he later raised to a postulate) that the speed of light was an Absolutely Unchanging Constant of Nature. So that even if one could move alongside a light beam at 99% the speed of light they would measure that light beam to be traveling 186,282 miles per second away from them, and not the 1,863 miles per second away from them as the Principle of the Addition of Velocities would naturally and normally require.

This then concludes the different interpretations when referring to the Constancy of Light's Velocity.

4

The Three Truths

$E=mc^2$, The Constancy of Light's Velocity, and The Expanding Universe

It is true indeed, that the speed of light is constant, that is, constant in the sense that it does in fact propagate from place to place with one specific velocity. However, I will attempt to show that the speed of light "Is Not," I repeat, Is Not an Absolutely Unchanging Constant of Nature.

In addition, I will attempt to show why Michelson and Morley always found the velocity of light to be the same regardless of the direction they pointed their light beams.

Finally, I will attempt to show why the Rest Energy, E, in Einstein's famous equation, $E=mc^2$, is always equal to mass, m, times the velocity of light squared, c^2.

The big questions then are: Why Does Light Not Have To Accelerate To Get Up To Speed? And secondly, Why Does It Travel At The Particular Speed That It Does (186,282 miles per second) and not say, 400,000 miles per second, nor 100,000 miles per second?

The answers to these questions can be found bound

up in *Three Basic Truths*—three facts of life which scientists have so diligently uncovered.

The first of these Three-Truths will be that which concerns Einstein's famous equation on mass and energy, $E=mc^2$.

What does $E=mc^2$ actually mean in plain and simple terms? It simply means that the "rest energy" of each and every piece of matter in the Universe is equal to the product of a body's mass and the velocity of light squared. Now I ask, how much less complicated can nature be? For what this actually means is that the rest energy which is bound up in everything and anything here on Earth (including the Earth itself) is equal to 90-Quadrillion (9.0×10^{16}) joules of energy for each and every kilogram of matter.

Take for example a lump of coal on Earth weighing 2.2 pounds. This is equal to a mass of approximately one kilogram. Or take a rock having this same mass of one kilogram; or a piece of metal of equal mass. Whichever material we use, it makes no difference. What $E=mc^2$ teaches is that the rest energy bound up in any of these one-kilogram chunks of material is equal to the same amount of energy required to electrically heat 1-million homes for one year (based on an average home using 25,000 kw hours per year for heating). This is derived at using Einstein's famous $E=mc^2$ simply by multiplying the coal's mass (or the rock or the metal's mass) of one kilogram by the velocity of light squared, which is about 300,000,000 meters per second squared. The product of this then is 9.0×10^{16} joules (a joule being a unit of energy). And since one kilowatt-hour is equal to 3,600,000 joules, we find there is an equivalent of 25-billion kilowatt hours of energy bound up in any one kilogram piece of matter in our piece of the Universe.

Hypothetically speaking, this enormous amount of rest energy should also be equivalent to what would happen if you could hit the Moon with a 2.2 lb. (one kilogram) rock in one and a quarter seconds from the Earth. Of course, the velocity of the rock would have to be an *unaccelerated* constant velocity of 186,282 miles per second from the very moment it left your hand (ridiculous as that sounds) to the very moment it hit the Moon. The resultant energy released upon impact would be tremendous. Theoretically, it too should be equal to the energy of 25-billion kilowatt hours of electricity.

This then is how nature can provide us with a "sensible and logical" explanation for such a phenomenal amount of stored energy bound up in such small amounts of matter.

There can only be one explanation for this phenomenon, and that is that the Earth, at its present position in space and time, is itself rushing off through space at a present speed of 186,282 miles per second, so that in effect, $E=mc^2$ is actually $E=mv^2$. In reality, this would mean that the rest energy equals the mass times *our own real and actual velocity squared*.

This being the case, we ourselves (including the coal, the rock, the Earth, etc.) are the real factors possessing the basic velocity which must be squared! Henceforth, both the propagation and the precise speed of light's propagation would follow as a direct consequence of the Earth's (or any other body's) basic velocity through space. Therefore, $E=mc^2$ lays itself bare to "reason" after all.

Einstein himself did not claim to "basically" know why his famous energy equation worked, but $E=mc^2$ did work, and it worked well.

Here then, in $E=mc^2$, lurks a piece of solid evidence in support of an uncomplicated model for the Universe, a piece of evidence, *which by itself,* should suffice to reveal to us the true secret of Light's Flight throughout the Universe!

The Second of the Three-Truths I refer to states that The Velocity Of Light Is Always Constant.

Now certainly, if it is indeed a fact of life that $E=mc^2$

is proof in itself that the Earth is rushing off through space at a present speed of 186,282 miles per second, it should then become crystal clear why the velocity of light (for all practical purposes) propagates with one specific velocity. I say, "for all practical purposes," because as I believe, light's velocity will be found to be steadily increasing by approximately 2.8 Angstroms per second per second as a result of that same increase in velocity of the body it is emitted from, that is, from the Earth and everything attached to it. The most amazing thing however about this *not so constant constancy of light's velocity,* is that when a light source is activated, such as, say a flashlight, its beam never starts out from zero-velocity as would logically be expected of anything which is presumed to be at rest, rather, its beam starts out *unaccelerated* with a velocity of 186,282 miles per second!

Now common sense and logic (contrary to those who hold that such niceties do not always prevail) tells us that *nothing, nor no one, can just up and go on a journey without having to start out!* Yet we know it to be an indisputable fact that a light beam does just that—it always begins its journey with a starting speed of 186,282 miles per second from a position which is "presumed" to be at rest. This is absolutely impossible in the realm of everything we know and see around us, yet that is precisely what light appears to do! Whatever happened to that part of its velocity range from zero through 186,281 miles per second? How did it manage to skip the natural stage of acceleration and just happen to begin its journey with a velocity of 186,282 miles per second? The only explanation which can realistically support the nature of this constancy of light's velocity is the same as that given as to why $E=mc^2$. And that is that the materials which the Earth and everything upon it are made (including the flashlight in this example) have been on a journey for billions of years (34-billion to be exact), constantly accelerating at 2.8 Angstroms per second per second, and are now presently traveling through space at the enormous speed of 186,282 miles per second. In other words, it is only logical to suspect that before switches are pulled, or buttons are pushed to activate a light source, that the light source itself (again the flashlight in this case) must already had been in motion, moving along with the Earth at 186,282 miles per second, and that the *unaccelerated* propagation of its beam of light at its one and only speed of 186,282 mps is simply a direct consequence of its present basic velocity through space in the first place.

In other words, the flashlight's cold (unenergized) image was *already* propagating at 186,282 mps; however, by energizing it we simply *enhance its image, making it hotter—and therefore brighter*—such that it then becomes readily visible when it impinges upon opaque materials. It would be analogous to a high velocity stream of water rushing past you. If you pour some red dye, for example, into the speeding stream, when the dye contacts the stream it initially has its velocity (without having to accelerate), i.e., if the stream's velocity is a hundred feet per second, the dye's initial unaccelerated velocity is also a hundred feet per second. This then is the case with the light beam from the flashlight. In effect, we changed the color (temperature) of the flashlight's filament in mid stream, i.e., the flashlight's *cooler* streaming image was already moving at c before altering the filament's image.

Here also, as with $E = mc^2$, Einstein did not claim to know why light propagated with one specific velocity. No one had ever explained why light propagated with one velocity, it remained a mystery.

However, it was this Constancy of the Speed of Light which played the major role in Einstein's Special Theory of Relativity. And though it was not actually he, but James Clerk Maxwell who first made it known that the speed of light always propagated with one specific velocity, Einstein made use of Maxwell's work entitled "The Nature of Light," and in 1905 followed up with his Special Theory of Relativity. In this, he indicated that not only does the speed of light always propagate with one specific velocity, but that its speed was also an Absolute and Unchanging Constant of Nature Throughout All Space and Time.

Here then, Einstein made a whole new assumption when he said the velocity of light throughout the Universe of space and time was an absolute and unchanging constant. All the usual predictions of his Special Theory of Relativity then followed, both, as a consequence of his four-dimensional Space Time Continuum, and his assumption that the velocity of light was the *maximum and limiting velocity*—the same unchanging velocity for all observers at all times—past, present, and future, throughout the Universe.

It was here then where it appears to be that the "monkey wrench" may have gotten thrown into the mechanism, because it was this *absolute and unchanging value of light's velocity* which thereafter became "A Way of Life." It became a widely accepted postulate of the Nature of Light throughout the scientific world, and therefore had most probably brought whatever real progress was being made to somewhat of a standstill.

If, on the other hand, it had been realized that the velocity of light was "Not" An Absolutely Unchanging Constant, but *Does In Fact Increase With Time*, it might then have been possible to more accurately answer questions such as: Will The Universe Go On Expanding Forever? Or, How can The Four Forces Of Nature Be Unified—specifically, Gravity, The Electromagnetic Interactions, The Strong Nuclear Force, And The Weak Nuclear Force?

Of course, if the Earth is in fact moving through space at a present velocity of 186,282 miles per second, and if indeed it is constantly accelerating at a rate of 2.8 Angstroms per second per second, then it should be relatively easy to calculate the magnitude of g at its surface.

Einstein did in fact teach that the effects of Acceleration

and Gravity are, for all practical purposes, *Indistinguishable* from one another. In that case, if the Earth is constantly accelerating at a rate of 2.8 Angstroms per second per second, then the magnitude of g (in units of Newtons per kilogram) at its surface, F, should strictly be due to the effects created from this extremely minuscule rate of acceleration by the simple equation ($F=m_r a_u$), where m_r represents a body's "Mass of One Square-Meter Radius" (A New Unit which reflects a body's Radius in *Cubic Meters* rather than its customary radius in *Meters*); and of course, a_u, which represents the Universal Constant of Acceleration.

In addition to Earth's minuscule rate of acceleration being responsible for creating its gravitational field, we should then know the Hubble Constant more precisely. This constant should be found to be in the range of 8–9 kilometers per second per million light years. Technically however, it will not appear constant much beyond 3–4 billion light years simply because "constant acceleration" can only be equated with a constant which is associated with *time,* and *not distance,* as the Hubble Constant connotes.

As for the issue which concerns the Deceleration Parameter (q_o), there should be no question that there is *No Slowing of the Expansion* if the galaxies are constantly accelerating at the rate of 2.8 Angstroms per second per second. This should instead be regarded as "Perfect Uniform Expansion" since the rate of expansion is constant for all time, throughout the whole of the Universe. It is strictly a *linear one* then when viewed as a *velocity-time* relationship, and *non-linear* when viewed as a *velocity-distance* relationship. For in any accelerating system of bodies where a strict velocity-time relationship exists, it will be found that the "Further Ahead In Time" we measure for recessional velocity between two bodies at *equidistant points,* the lesser will their recessional velocities be as they increase their distance *per unit time.* As a result, it is plain to see that recessional velocities *per unit distance* will decrease as time goes on, thereby imparting a somewhat false sense of a *deceleration effect* when looked at from a velocity-distance point of view.

In contrast to this, if we were able to view this expansion process of galaxies in the opposite direction (that is in the direction of t=0) we should find as we look further and further behind us, so to speak, that there exists just the opposite effect: the galaxies (per unit distance) recede from each other faster and faster as we look further and further back towards the center (towards the younger galaxies) where the expansion process supposedly began. This viewpoint would impart a "false sense" of a Speeding Up Of The Expansion, when in fact it would really be an indicator of "true uniform expansion." What we then have, is that when we detect deceleration ahead of us, that is, away from t=0 in the expanding system per Unit Distance; and acceleration behind us (in the direction of t=0) in the expanding system per Unit Distance, these are only an assurance that strict uniform expansion is indeed taking place. In other words, there is *no slowing whatsoever* of the acceleration process going on in the Universe. It is indeed expanding, but the fact is that there can be no "real" Hubble Constant in relation to "velocity and distance" over the long run; the real constant can only apply to a "velocity-time" relationship, as is depicted in Figure 1.1, The Universe In Numbers.

In this, the third of The Three-Truths, the Universe is in a state of Uniform Expansion. Although it was Slipher who showed that the galaxies are receding from us, it was through the very meticulous work of the American Astronomer, Edwin Hubble, that the secret of uniform expansion was revealed.

Hubble's discovery centered on the fact that the distant clusters of galaxies were receding (moving away) from us and each other at velocities which appeared to him to be directly proportional to their distance. Hubble arrived at this conclusion for the simple reason that when he found, for example, a cluster of galaxies to be a million light-years away, their recessional velocity was determined by him to be approximately 100 miles per second; and when he found galactic clusters 2 million light-years away (twice the distance as the first) their recessional velocity was found to be 200 miles per second, twice that of the first. If he detected galaxies at a distance of 20 million light-years, they were measured to be receding at 2,000 miles per second. (Note: Velocity-Distance Scale is now estimated to be only about 10% of Hubble's Original Estimate.)

Hubble supposedly found this same pattern no matter how far out in space he looked. He consistently found that the recessional velocity was apparently in direct proportion to distance, and therefore took this to mean that the Universe was expanding in a uniform manner.

If we compare this uniform expansion, for example, with recessional velocities and expansion rates between solid bodies falling to Earth, it will be found that, except for the fact that all galaxies *diverge* from each other, and a multiplicity of falling bodies on Earth *converge* and fall towards a common center, there is not the slightest difference in the structure of the equations describing their motion, except for their particular constants of acceleration of 9.8 meters per second per second for Earth bound objects, and 2.8×10^{-10} meter per second per second for space bound galaxies.

The method Hubble used to make his far reaching determinations comprised the use of the well known Doppler Effect. With the Doppler Effect, the wave length of light is shifted toward the blue end of the spectrum (high frequency, shorter wavelength) when a source of light is moving towards an observer. When a source of light is moving away from an observer, the wavelength of the light is shifted towards the red end of the spectrum (lower frequency, longer wavelength) in which the waves appear to be spread out. Also, with the Doppler Effect, the greater the speed, the greater the shift in wavelength. Therefore, by determining the extent of displacement in the spectral lines, it then becomes relatively easy to determine the recessional velocity between the observer and the source. Once this recessional velocity is known, it then becomes a matter of simple mathematics to estimate the distance of galaxies, since the Velocity-Distance relationship appeared to Hubble to be in direct proportion to each.

What does this expansion process mean? First, I might comment that Hubble's discovery that The Universe Is Expanding In An Orderly Fashion was the most important Cosmological Discovery ever made!

Originally, Hubble estimated the recessional velocity between galactic clusters to be about 100 miles per second per million light-years (our present light-year is about 5.9 trillion miles—the distance traversed by light at the rate of 186,282 miles per second during a time span of one year). However, after many years of investigation of the Velocity-Distance Scale by others, the constant has now been revised all the way down to approximately 11 miles per second per million light-years, *although it is still disputed.* This latest refinement of the Hubble Constant is the result of many years of work by astronomers such as the distinguished Allan Sandage. It is a number which I have found to be in excellent agreement with my own equation for determining the Universal Constant of Acceleration.

Now although the accuracy of the Hubble Constant is extremely critical—both for the purpose of determining a_u, and for the purpose of determining, more precisely, the age of our own galaxy—the real significance of Hubble's discovery is the fact that the Universe of Galaxies are indeed expanding uniformly, that is, they are spreading apart from each other in an orderly fashion which is commonly referred to as the Hubble Law. Hubble's Law simply states that the further away a galaxy is, the faster it is receding from us.

In order to understand what is actually taking place in a Uniformly Expanding Universe, I can think of no better way to describe it than to relate it to something we are all familiar with here on Earth—and that is the influence which gravity has upon bodies in free fall as relates to rate of fall and distance traveled. In fact, it will be seen that the same physical laws which apply to freely falling bodies under the influence of gravity here on Earth, are precisely the same physical laws which apply to the Expansion of the Universe.

The only differences to be found between these two expanding systems will be that one is a *convergent* system, while the other is a *divergent* system. And other than their rates of acceleration being radically different, they both obey the same physical laws. Let us use for example a quantity of golf balls held at rest about 5,000 feet above the Earth's surface, then release them. What happens naturally, is that they would all move from their positions of rest and converge towards the center of mass of the Earth as they accelerate at a constant rate of speed of about 32 feet (9.8 meters) per second per second near the Earth's surface. The only difference between this system and the *Expanding Universe of Galaxies* is that the galaxies all move outward, *diverging* instead from each other through space, (rather than converging as with gravity on Earth) and accelerating at a constant rate of 2.8 Angstroms per second per second (about one ninety millionth of an inch per second) as in contrast to 32 feet per second per second for the converging golf balls. So that when the galaxies go forward at their constant rate of acceleration of 2.8×10^{-10} m/s^2, they will attain a speed of 186,282 miles per second at the end of 34 billion years, just as our own galaxy has, and will have traversed a distance from t=0 *in one general direction through space* of 17 billion of our present light-year of 5.9 trillion miles; whereas, when the golf balls fall toward the Earth's surface at a Constant Rate of Acceleration of 32 feet per second per second, they will be traveling at a speed of 320 feet per second at the end of 10 seconds, and will have traveled a distance of 1,600 feet.

To elaborate further on these two systems, let us say that the whole Earth is literally covered over by a dozen or so layers of golf balls situated about a mile above its surface. Each layer is spaced, say ten feet above the other, and the golf balls in each layer are spaced about 300 feet apart. Now then, all the balls in the first layer surrounding the Earth are released simultaneously, and in each ensuing second of time another layer of balls are released until in twelve seconds time all twelve layers of balls are released.

Let us also suppose, for the sake of argument, that all the balls continue to accelerate at a constant rate of 32 feet per second per second (ignoring the fact that the Earth is a solid body) and they fall 4,000 miles in 19 minutes to the very center of mass of the Earth where they will impact with a final velocity of 7 miles per second—converging into a somewhat hot solid mass.

As can be seen, the only differences between the two systems are that the golf balls accelerate inwardly to eventually converge upon each other at a rate of 32 feet per second per second, while the galaxies accelerate outwardly *in a divergent fashion* through an infinite space away from each other at only 2.8 Angstroms per second per second.

Of course, there is no comparison in their respective rates of acceleration since the golf ball's acceleration rate of 9.8 meters per second per second on Earth is approximately 35 billion times higher than the acceleration rate of the Earth and Galaxies through space of only 2.8×10^{-10} meter per second per second.

In each of these situations however, *both*, the golf balls, and the galaxies, *double their velocity at four times the distance traveled.*

As can be seen then, there is *no difference* whatsoever in the principal laws which govern the motions of accelerating bodies—whether they are galaxies accelerating through infinite space in a divergent fashion, or golf balls accelerating to Earth through a finite space in a convergent fashion. In any case, each system accelerates in strict accordance with the same basic laws of nature.

5

"Why" the Velocity of Light is *Not* Constant

In the last chapter I had spoken mostly in general terms about the three most tested, most accepted, and most important cosmological discoveries of our time. These are: Einstein's Energy Equation, $E=mc^2$; The Constancy of the Velocity of Light; and the discovery of a Uniformly Expanding Universe—uniform in the sense that it is a linear type velocity-distance expansion. I of course advocate a *non-linear* velocity-distance expansion since I have stated previously the strict proportionality of a velocity-time relationship.

It is my opinion that each of these three discoveries can stand alone to support what I have stated over and over—that is—that it is we ourselves, along with the Earth and the rest of the galaxy, that are accelerating through space at a present velocity of 186,282 mps; and that both the propagation and the precise speed of light are direct consequences of our own basic motion through space.

Why shouldn't this conclusion have been reached a long time ago? The answer to this is not a simple one, and may not even be a nice one. I say this only because if what has been concluded in this book is indeed correct about the basic mechanism of light, then it may have to bear upon Einstein's Theory of Relativity, for although Einstein contributed many good and positive works in addition to his energy equation, it was however, he who solidly established the postulate that *light propagated with one absolute and unchanging velocity,* i.e., that its speed was one and the same speed in the past, is the same at present, and will forever remain the same in the future—not only here in our vicinity of space—but throughout the whole of the Universe.

Now certainly, it is not debatable to say that light normally propagates at one specific speed. However, if we instead say that light propagates with only one *absolute and unchanging* speed throughout all space and time (as Einstein had postulated) it then becomes a debatable issue! Still however, there is nothing wrong if we say light propagates with "one constant velocity," that is to say, "constant for all practical purposes"; for if the only change in its velocity of 186,282 miles per second amounts to an increase of, not kilometers per second, nor meters per second, nor even centimeters or millimeters per second, but only approximately Three-Angstroms, or one ninety-millionth of an inch per second, then we can indeed state for all practical purposes that light does propagate from place to place with a constant and or unvarying velocity.

So there we have it, the velocity of light is, *for all practical purposes,* constant, just as Einstein and Maxwell before him taught. However, Einstein may have overlooked one very important characteristic about light which he must not had realized, and that is its very small, very minuscule and constant increase in its speed by approximately 2.8 Angstroms per second per second. Granted, that is only an increase in speed for each second of time of *one part*

"Why" the Velocity of Light is *Not* Constant   35

in a thousand-quadrillion of its present speed; but it is in fact an increase, insignificant as it may seem.

Indeed, it is this small but constant increase which is the only factor in the expansion of the Universe. The propagation of light itself appears then to be a secondary effect arising out of the basic motion of the material of which galaxies are made—the electrons, protons, and neutrons, etc.

As for Gravity, there should be no question of its basic mechanism once it is fully realized that the massive Earth, for instance, is moving through space with a constant acceleration of 2.8 Angstroms per second per second, just as every other body in the Universe is doing.

As stated earlier, Einstein showed that the effects of acceleration and gravitation are *indistinguishable* from each other. Therefore, if indeed the Earth is accelerating, we should confidently be able to state that the force of gravity on Earth is solely attributable to the fact that it is presently rushing through space with a constant acceleration of approximately 2.8×10^{-10} meters per second per second.

It should then be a matter of simple mathematics to determine why, for example, an object weighs what it does on Earth, or why that same object weighs only one-sixth Earth's value on the Moon, since naturally, whatever the Earths acceleration rate and velocity through space are, so too is the Moons'.

As for finding a direct relationship between the Forces of Gravity and Electromagnetic Interactions, it would seem evident then, *what that connection is!* It is simply that the gravitational force should be due to the Earths acceleration through space at a constant rate of 2.8×10^{-10} meter per second per second; and that the precise velocity of light should then be due to the Earths present velocity through space of 186,282 miles per second after steadily accelerating for 10^{18} seconds, otherwise, 34 billion years.

I am not aware of any existing technology which affords us the capability of detecting an increase in the velocity of light by as little as *three Angstroms per second* (one centimeter per year). If such technology could afford us this capability, then of course we would be in an excellent position to test this theory which, in effect, would be a direct method to determine both, our basic velocity through space, and our basic acceleration rate through space by *using light itself as the measuring rod.*

This of course is what Einstein's Special Theory of Relativity runs directly contrary to, whereas he himself had decreed that "It is futile to try to determine the true velocity of any system by using light as a measuring rod."

So once again, I repeat: It Is This Absolute And Unchanging Value Of The Speed Of Light *which Einstein decreed to be a fact of life* which I feel has *slowed progress* in the fields of both Astrophysics and Quantum Physics.

I have wondered for many years now why, with so much solid evidence as these Three-Truths confront us with, why we had not come to the conclusion a long time ago *that the basic motion which determines the speed of light is the same basic motion of the planets and stars.*

I suppose the front runner of any argument to the question might well be that which Einstein himself had postulated when he said the velocity of light is an absolute and unchanging constant of nature throughout all space and time.

Now it is true that light does indeed propagate at only one speed! But what this means is that light does not have to go through the process of accelerating from zero-velocity, i.e., from rest, to get up to its present speed of 186,282 miles per second. We can safely state that much about light since that much about the character of light is proven beyond any doubt. Please overlook, if you can, the repetition in making my point. It is simply *too important* not to be repetitive.

To go on then, think about this if you will: think about

"Why" the Velocity of Light is *Not* Constant

the way in which light propagates at the colossal speed of 186,282 miles per second and *Never Even Has To Start Out To Attain That Speed*. Ponder upon this for awhile and then ask yourself: How can anything in nature just simply up and take off with an initial speed of 186,282 miles per second from a position of presumed rest *without first having to accelerate* to get to that speed? First of all, it must be realized that one factor is clear and indisputable—and that is that light does indeed always propagate in empty space with only one constant velocity, 186,282 mps at the present time, however little it may increase from second to second.

Now then, since there is no dispute whatsoever of that one particular characteristic of light, there should then be no argument that the source which emits the light must itself had *already been in motion at that particular speed,* and for that reason, and that reason alone, light's built in instantaneous velocity is what it is.

I believe that when the idea first emerged of the "Constancy of Light's Velocity" from Maxwell's equations, it was only intended to be construed that it did not have to accelerate up through the velocity range from zero to 186,282 mps to get to 186,282 mps, but instead always propagated at one instantaneous built-in speed for whatever the reasons. Einstein then concluded that this constancy of the velocity of Light should also represent that its velocity never *throughout all space and time*, changed even infinitesimally. And so, to my knowledge, it was Einstein who put the real emphasis on the velocity of light being a limiting velocity—an absolutely unchanging velocity for all time and space throughout the Universe.

Ironically, it was Einstein himself who *may have* put the one obstacle in his path during the last 30-years of his life which he spent trying to Unify Gravity and Electromagnetism *but failed*.

6

"Why" $E=mc^2$

As I have contended all along, we certainly must already be in motion through space, racing along with a present speed of 186,282 mps in order for light and all other forms of electromagnetic radiation to leave their source with a *built-in starting speed of 186,282 mps.*

I do not feel we should need any additional evidence to convince us of our actual flight through space at the speed we attribute to light. However, I take up once more the important discovery which Einstein himself made concerning Mass and Energy. No one can deny the validity of Einstein's equation on mass and energy, $E=mc^2$. In the previous sections I have explained what this equation meant. You will recall that I used as an example a lump of coal, a rock, or a piece of steel, each having a mass of one kilogram, and illustrated how and why, according to Einstein's famous equation on mass and energy, they each possess their enormous stores of energy.

Now no one, including Einstein himself, knew precisely why such a phenomenal store of energy resided in all material bodies attached to the Earth, including the Earth itself.

Imagine what this so called rest energy would be for the Earth as a whole? To envision this, consider that the Earth is rushing through space *as I propose it to be,* at a speed of 186,282 mps. Let us say that it is heading for the Sun, and the Sun is instead a cold dark star having a somewhat similar density and or material make-up as the Earth. And let us also say the Sun is not moving, that it is just suspended there in space. Now imagine the Earth running head on into the Sun at a speed of 186,282 mps. The total release of energy at that moment would actually be equal to the exact amount of energy which $E=mc^2$ predicts the Earth presently possesses—about 5.4×10^{41} joules. The same applies to the Sun, and according to $E=mc^2$ its rest energy amounts to 1.8×10^{47} joules. However, this stored or rest mass would be exactly equal to that which would be released if the Sun instead were to run head on into a stationary solid body (similar to the Sun itself) at a speed of 186,282 mps.

Also, imagine if you will, two hot stars such as the Sun, each rushing towards each other from the outer edges of two different expanding Universes. Each then collides with the other head on at a speed of say 300,000 kilometers per second, if not more. The energy and total light output from two hot stars such as the Sun colliding head on at 300,000 km per second each, should in theory be equal to what is happening when we are looking at Quasars, for if what I have proposed in this book are facts to be reckoned with, then there may just be good reason to believe that Quasars are simply *two colliding stars near the outer edges of our own expanding Universe and another expanding universe, both of whose outer edges have eventually come to expand into each others turf.*

"Why" $E=mc^2$ 41

Getting back however to the philosophy of $E=mc^2$, as I have pointed out over and over, $E=mc^2$ is one of the Three-Truths I refer to. The equation itself is not at all in dispute. Einstein had given to the world a great contribution in $E=mc^2$. So I ask of the reader, once again, to ponder upon this famous equation. Ponder its implications, consider how significant it has been to science, and how it stands today upon solid ground as to its validity.

Now since no one has yet explained precisely why $E=mc^2$ works, and since there is no dispute whatsoever that light does indeed propagate at one speed; and considering my own reason why light only propagates with one speed, how else then can a mass of only one kilogram store such an enormous amount of energy, *energy sufficient to heat a million homes for one year,* if it were not for the plain and simple fact that *the one-kilogram mass is itself speeding along with the Earth through space at the phenomenal speed of 186,282 mps?*

This then seems like the only rational explanation why Einstein's energy equation calls for multiplying the mass by the velocity of light squared. It is, in reality, the velocity of the coal, rock, steel, etc., which is actually being squared; so that in essence, *the equation is really $E=mv^2$.*

So unless we decide that common sense and logic should be discarded, the velocity of light squared in Einstein's famous equation ($E=mc^2$) seems clearly to stand for the actual squared velocity of the object of mass in question; and so, the energy (E) would in effect actually be *Kinetic Energy!* And in this case the object is a rock, a lump of coal, an automobile, or the Earth itself, all of which have a basic velocity of 186,282 mps by virtue of their basic velocity through space at their present space-time position.

This then, very simply, is how I believe Nature has managed to give to light waves, radio waves, and all forms of electromagnetic phenomena, a vehicle with which to move from place to place. As such, Einstein's Principle of *The Addition of Velocities* would apply to light after all. However, our measurements may never provide us with anything except one observable velocity for all observers. See for example, figures 1–3 through 1–7.

7

"Why" the Universe Expands

So much said for $E=mc^2$ and the Constancy of the Velocity of Light. If these two truths should not suffice to convince one that they are traveling along with the Earth at the speed we attribute to light, then Hubble's discovery of a Uniformly Expanding Universe should provide a third piece of evidence.

The most significant factor about the Expanding Universe centers upon Hubble's Constant. This constant provides us with, not only the rate at which the Universe expands, but put another way, the "Rate of Universal Acceleration Through Space" of all matter throughout the Universe.

In other words, since the present estimated value of the Hubble Constant is approximately 18 kilometers per second per million light years, then it is only reasonable to work this figure backwards, that is, closer in to our own galaxy, all the way in to a distance from Earth of only *one light-second*, 186,282 miles. When this then is finally reduced from 18 kilometers per second per million light years, all the way down to only one light-second, the resultant figure is about 5.6×10^{-10} meter per second per light-second, otherwise 5.6 Angstroms per second per light-second. In this manner, we can presume that the Earth and all other matter in the Universe is accelerating through space at a constant rate of 5.6 Angstroms per second per second.

As I indicated throughout the book however, the actual near field constant should eventually be found to be about 9 kilometers per second per million light years (instead of 18 kilometers) since the velocity of light is *"not"* after all constant. This of course will then fit in with a Universal Constant of Acceleration of approximately 2.8×10^{-10} m/s^2 which is the actual rate required to correspond to the magnitude of g at the surface of all material bodies throughout the Universe.

Consequently, what we presently believe to see at 1 million light years away should actually prove to be 2 million light years away, and so on. Therefore, 2.8 Angstroms would then be appropriate, as this would reflect a *real* Hubble Constant of 5.5 miles per second per million light years—or—28.7 km per second per megaparsec instead of 57.4 km per second per megaparsec.

Hubble's Universe is one of simplicity and beauty, however, we seem to have found a way to complicate the picture. Let me explain why. If one were to indulge in, say fifty, or even a hundred books related to cosmology, it would easily be concluded that the Universe is basically very simple in its structure—and rightly so.

Hubble's discovery revealed that not only was the Universe expanding, but that all galactic clusters were moving away from each other at speeds which were in direct proportion to their distance from each.

Now no matter which of the literature one indulges

in on this subject they will find that all or most give the same interpretation of the Expanding Universe. This interpretation generally states that all clusters of galaxies in the Universe are moving away from each other. They will generally teach—and rightly so—that as time goes by, all galactic clusters move further and further apart, and that this expansion process has been going on for close to 20-billion years (my estimate is 34-billion years for our own space-time epoch). They will also indicate that the size of the Universe back about 20 billion years ago when it first began to grow, was extremely small, and at that particular time in the past when the Universe was just beginning to take form, all the material which the galaxies were formed from was crowded together. It is also generally stated throughout the literature that a large explosion took place (commonly referred to as the Big Bang), and as such, the material of this infant Universe flew apart in all directions, so that at the present time, the distances between all clusters of galaxies are much greater than they were in the past, and will continue to increase at a constant rate.

As you may see by now, if there is one statement that seems clearly remarkable when attempting to describe the Universe, it is simply that all galactic clusters are "spreading apart from each other."

It is advocated however, "that it is not actually the galaxies themselves" that are flying off into outer space, but rather, it is *space itself which is expanding*.

Now there is simply no question that the galactic clusters are moving away from each other. That is no hypothesis, it is happening indeed!

The argument however which favors an expanding space is itself a very ambiguous one, if not a very weak one. I am not sure whether it is even clear to anyone what an expanding space means, for then, *what is there where the space did not yet expand into?* The idea in itself only seems to create additional problems in attempting to model the Universe, so I will let it drop here for what it is worth.

Back then to the real issue. Now I ask, is it real that we are observing galactic clusters moving away from us at a rate of speed of what appears to be approximately 11 miles per second for every million light-years distant from us! Yes, of course it is real to believe this because it is indeed a fact, an actuality, and a truth, and what's more, it has all been found very objectively!

This expansion then is actually a true indicator of how fast the galaxies are moving through space, and above all, it is a true indicator of the rate at which we and other galaxies *are accelerating through space.*

Is this expansion not then a Universal Constant of Acceleration—whether it is stated as 5.5 miles per second per million light years, or 2.8 Angstroms per second per light second?

I will answer my own question and say to you, yes! "It Is Real Indeed" to believe that *The Expansion Rate Is Strictly The Universal Acceleration Rate*—that is all it can be! It is, for example, the same precise pattern we would find if we could be intelligent microbes riding upon one of many golf balls falling near the Earth's surface. And except for the difference in direction and the large difference in the ball's rate of acceleration, *32 fps/ps instead of 2.8 Angstroms per second per second*, there are no other differences whatsoever. The same basic equation—the same law of nature which applies to a freely falling body on Earth—also appears to apply to the outrushing galaxies.

Now if we will agree that the Hubble expansion rate (on the basis of one light-second) is, in reality, the rate of acceleration of all material bodies in the Universe, then it is also real to say that the Earth and everything attached to it are accelerating through space at a constant rate of 2.8 Angstroms per second per second.

So where do we go from here? What else do we know

"Why" the Universe Expands 47

as reality? Well, we know it is real that the velocity of light is constant, that is, it propagates at one speed when it leaves its sources without having to accelerate to get to that speed. From this we can reason that our own velocity through space must then be that which we attribute to light, (since light already possesses this built-in velocity) for how else can light possess such a quality if it were not for the plain and simple fact that it's source *is already moving* at such a phenomenal speed in the first place.

This velocity turns out to be precisely the velocity of our own planet through space if we say the materials our Earth are made of began their journey about 34 billion years ago. For if we then multiply the number of seconds of time in 34 billion years by 2.8 Angstroms we will come up with approximately 186,282 miles per second.

In other words, a body such as planet Earth accelerating for approximately 34 billion years at a constant rate of 2.8 Angstroms per second per second, will have a velocity of 186,282 miles per second at the end of the 34 billion year period (34 billion years multiplied by 31.6 million seconds equals approximately 1,100 Quadrillion seconds of time). Multiply this product by the 2.8 Angstrom constant rate of acceleration and it equals approximately 3.0×10^{18} Angstroms (300,000 kilometers) per second at our present space-time position relative to the original location in space where the material of our galaxy first began to accelerate from. I will more or less *humorously* call it the vicinity of the "Big Began" since I do not feel the Big Bang Theory is the correct one. I believe the Universe had instead, a long beginning in the makings of its birth, a very long and very slow beginning, starting out at 2.8 Angstroms per second in which it was *not hot* at all, but instead *very cool,* if not cold, by our standards. There were probably no high temperatures which came into existence until millions, or even billions of years have passed in order for velocities to build up sufficiently for particles to agglomerate and eventually heat up into large volumes of hot gaseous bodies.

We should therefore look upon the birth of the Universe and its expansion in somewhat the same manner as we look upon all other forms of "Biological Processes" here on Earth; for if there is one thing we surely know of the biological growth process, it is that it always begins in a non-explosive manner, i.e., with just one or two or a few small particles of matter (never ever exploding instantaneously into being) but rather, very, very gradually developing and growing.

This I believe is really the manner in which the Universe itself has grown to what it is at present—*Not Out Of A Big Bang At All*—but more in the spirit of *a very long beginning,* which as stated before, might be more appropriately (if not humorously) ascribed to as the "Big Began"!

8

The Universe in Numbers

The picture of The Universe In Numbers (See Fig. 1–1) may help explain some of the enigmas which presently exist in the field of cosmology, especially in relation to two very important parameters: the Deceleration parameter, q_o, and Hubble's constant, H_o.

The numbers picture (Fig. 1–1) shows the Hubble Constant to be 5,472 miles per second per billion years. Note however, that Fig. 1–1 reflects a "velocity-time" relationship and not the "velocity-distance" relationship customarily associated with Hubble's Constant. Broken down even further, into the million year increment, the constant would then be 5.472 miles per second per million years.

What is of significance here is that this number applies to *time* instead of *distance* and is approximately *half the value* of the present estimate of the Hubble Constant.

A third interesting point is in reference to our position in space-time (relative to time-zero, $t=0$, where it is gener-

49

2.8 Angstroms

AGE (BY's)	VELOCITY (mps)	GALACTIC CLUSTERS (BLY's)		LIGHT YEARS
1	5,472			14,687,366
2	10,944			58,749,462
3	16,416			132,186,290
4	21,888			234,997,849
5	27,360			367,184,140
6	32,832			528,745,162
7	38,304			719,680,915
8	43,776	1		939,991,399
9	49,248			1,189,676,614
10	54,720			1,468,736,561
11	60,192			1,777,171,239
12	65,664	2		2,114,980,648
13	71,136			2,482,164,788
14	76,608	3		2,878,723,660
15	82,080			3,304,657,262
16	87,552			3,759,968,596
17	93,024	4		4,244,648,661
18	98,496			4,758,706,458
19	103,968	5		5,302,138,985
20	109,440			5,874,946,244
21	114,912	6		6,477,128,234
22	120,384	7		7,108,684,956
23	125,856	8		7,769,616,408
24	131,328			8,459,922,592
25	136,800	9		9,179,603,507
26	142,272	10		9,923,659,153
27	147,744	11		10,707,089,530
28	153,216	12		11,514,894,439
29	158,688			12,352,074,470
30	164,160	13		13,218,629,047
31	169,632	14		14,114,558,352
32	175,104	15		15,039,862,385
33	180,576	16		15,994,511,499
34	186,048 (Local Group)	17	(MW Galaxy)	16,978,594,646
35	191,520	18		17,992,022,873
36	196,992	19		19,034,825,831
37	202,464	20		20,107,003,521
38	207,936	21		21,208,555,942
39	213,408	22		22,339,483,094
40	218,880	23, 24		23,499,784,977
41	224,352	25		24,689,461,591
42	229,824	26		25,908,512,937
43	235,296	27		27,156,939,014
44	240,768	28, 29		28,434,739,822
45	246,240	30		29,741,915,361
46	251,712			31,078,465,632
47	257,184	32		32,483,412,084
48	262,656	34		33,839,690,367
49	268,128	36		35,264,364,831
50	273,600			36,718,414,026
51	279,072	38		38,201,837,953

Fig. 1–1. THE UNIVERSE IN NUMBERS

The Universe in Numbers

AGE (BY's)	VELOCITY (mps)	GALACTIC CLUSTERS (BLY's)	LIGHT YEARS
52	284,544	40	39,714,636,611
53	290,016	42	41,256,810,000
54	295,488		42,828,858,121
55	300,960	44	44,429,280,972
56	306,432	46	46,059,578,555
57	311,904	48	47,719,250,869
58	317,376	50	49,408,297,914
59	322,848	52	51,126,719,690
60	328,320		52,874,516,198
61	333,792	54	54,651,687,437
62	339,264	56	56,458,233,407
63	344,736	58	58,294,154,108
64	350,208	60	60,159,449,541
65	355,680	62	62,054,119,705
66	361,152	64	63,978,164,599
67	366,624	66	65,931,584,226
68	372,096	68	67,914,378,583
69	377,568	70	69,926,547,672
70	383,040	72	71,968,091,492
71	388,512	74	74,039,010,043
72	393,984	76	76,135,498,453
73	399,456	78	78,268,971,339
74	404,928	80	80,428,014,084
75	410,400	82	82,616,431,560
76	415,872	84, 86	84,834,223,767

Fig. 1–1 (con't.)

ally believed the Center of Mass of the Universe originated). It is no coincidence that our local group of galaxies which the Milky Way belongs to, just happens to be at the unique position in the Universe where our travel time in years (34 billion) is exactly twice our distance in light years (17 billion) from time-zero. This is simply due to the fact that our own present light year is naturally being used as the standard light year. Therefore, any galaxy throughout the Universe will naturally always be twice its years in age as its number of light years in distance when using a particular galaxy's own present light year as the standard measure. Otherwise, since the velocity of a galaxy increases with time, the age should then match the distance in light years, but of course, it would not be practical to use this equal system of years and light years—time and distance. So much then for why our distance in light years from time-zero is exactly half the number of our age in years.

What The Universe In Numbers indicates however is that the Milky Way (at 17 billion of our present 6 trillion mile light year from time-zero) has increased its velocity an average of 10.944 miles per second for "every million of our present 6 trillion mile light years of distance" it has traveled through space. Or it has increased its velocity by 10.944 miles per second for each *two million year period of time* it has traveled. So that we may state that the Hubble Constant should really be equal to 5.472 miles per second per million years of *time*—exactly half its present estimated value per million light years.

Notice I did not use light years with my own estimate. I used *years* because, according to the numbers picture in Fig. 1-1, the constant applies strictly to the *travel time* through space by a galactic cluster, and *not distance* as is customarily applied to the Hubble Constant.

One can see however, that in the vicinity of 3 to 5 billion light years in either direction of the Local Group (which includes the Milky Way) it would be easy to draw the conclusion that a galaxy's recessional velocity is almost strictly proportional to its *distance* from us. And as can be seen in any other galaxy's vicinity of 3 to 5 billion light years, a hypothetical observer on any of these galaxies, (especially from 10 billion light years outward from time-zero) would likewise have good reason to suspect that recessional velocity and distance had a strict linear relationship as the Hubble Law connotes, when in fact, it does not, at least not in the long run. For as the numbers picture in Fig. 1-1 depicts, even though the velocities of all galaxies constantly increase by 5,472 miles per second per billion years (5.472 miles per second per million years—or—2.8 Angstroms per second per second) their recessional velocities *per unit distance* continually decrease as they travel further and further, faster and faster, from their original position of time-zero.

Of course, if we look at recessional velocities of those galaxies that are traveling slower through space than ours (the *inward and younger galaxies* which are naturally closer to time-zero than the Milky Way) we should then find their recessional velocities *per unit distance* to be *increasing* as we look further and further towards t=0, and might therefore mistakenly believe the expansion process to be speeding up (overly accelerating instead of decelerating), when in fact it is absolutely constant in relation to time, just as applies towards the outward or opposite direction from us, i.e., the outer edge.

What the picture of The Universe In Numbers also indicates, is that when we look toward the outer edges of our own material Universe and detect large Red Shifts that equate with recessional velocities close to our own speed of light of 186,000 miles per second, it may just be that we are detecting galaxies that are (not 17 billion light years or so distant from us as is generally believed when red shifts indicate recessional velocities of 186,000 miles per second) but rather, galaxies that are 50 to 51 billion of *our* light years away instead, for that is the distance which Fig. 1–1 on the right hand column depicts between our local group of galaxies at the 34 billion year epoch, and galaxies if any, at the 68 billion year epoch. The recessional velocity of 186,000 miles per second would be the difference between our present velocity of 186,000 miles per second (at our 34 billion year, 17 billion light year mark in space-time) and the velocity of the hypothetical galactic cluster moving at 372,000 miles per second at its 68 billion year (also 68 billion light year) mark. We must realize here that at the 68 billion year mark in space-time, a light year is approximately *12 trillion miles instead of 6 trillion miles,* as our own present light year is.

In our first 34 billion year period of time we have traveled 17 billion of our present 6 trillion mile light years from time-zero, and in a second and equal period of time of 34 billion years, we will then have traveled (not another 17 billion of our present standard light years) but 51 billion of our present 6 trillion mile light years to the 68 billion light year (also 68 billion year) mark from time-zero. At this point in space-time we would then normally use 12 trillion miles as the standard light year since we would be moving at 372,000 miles per second. We would therefore determine by this gauge that our galaxy was 68 billion years old, and 34 billion light years distant from time-zero (t=0). I emphasize this is 34 billion light years of the *12 trillion mile light year.*

In contrast to this, whenever we penetrate with our telescopes deep in the direction of time-zero and detect some of the younger material which galaxies stem from—and of course galaxies themselves—we should again detect red shifts which indicate recessional velocities of 186,000 miles per second. However, because we are now observing the slower moving galaxies which began their journeys through space after ours, we should be seeing them at a maximum distance of only 17 billion of our present light year instead of the 50 to 51 billion light years which we should see of galaxies moving in the outward, or opposite direction from our position in space.

There are considerable differences however about detecting galaxies and other nebulous matter of the Universe when we attempt to penetrate inwardly (that is, in the direction of time-zero) as in contrast to looking toward the outer edge. I speak of this situation only in a hypothetical sense, for since we ourselves are at the 34 billion year, 17 billion light year mark from time-zero, we cannot expect to observe younger images of any material matter which is actually located much further back than our 8.5 billion light year mark, which is also the 24 billion year mark in time from t=0. In other words, from our present space-time position we should be able to detect younger images of galaxies which are actually located at a distance of approximately 8.5 billion of our present 6 trillion mile light year—galaxies which have an age of approximately 24 billion years and a velocity through space of about 132,000 miles per second (that too, of course, is their present speed of light).

The other difference in peering inwardly toward t=0 would be that at these distances and ages, some of the slower propagating material of the Universe may not have yet grown or formed into mature galaxies of any distinction. In any case, any images of material matter we might possibly detect that is 8.5 billion of our light years distant toward time-zero, would have to be of the 2-c radiation of that material's birth at time-zero (see Fig. 1–9, 2-c images). Actually, at approximately that point in space (8.5 billion light years from time-zero, or half our distance towards time-zero) we should hypothetically be able to detect the makings of the birth process in which galaxies would eventually form—the very beginning! Realistically however, *we would see nothing,* so to speak. Of course, closer in to us, *in the direction of time-zero,* we should naturally expect to see more and more, as is presently the case of nebulous matter and maturing galaxies, since they have had more time to develop.

This then is most likely the reason for the beliefs by some astronomers that the expansion process of the Universe of galaxies is slowing.

On the surface it would certainly appear that *deceleration* was taking place when we are observing older galaxies than ours, but then, we must begin to see it for what it really is: a strict and positive *accelerating system* where all the physical laws of nature which apply to freely falling bodies on Earth, for example, are precisely the same laws which apply to the Expanding Universe—except of course in their numbers—wherein g equals approximately 9.8 meters per second per second on Earth, and Hubble's Constant, H_o, *which I also refer to here as the Universal Constant of Acceleration,* a_u, equals 2.8×10^{-10} meter per second per second, otherwise, 5.472 miles per second per million *Years.* Once again however, I emphasize that in the vicinity of about 3 to 5 billion of our own present light years in any direction from the Milky Way, a recessional velocity of approximately 5.472 miles per second per million "light years" would in fact be apparent.

In the table which depicts The Universe In Numbers (see once again Fig. 1–1) the instant picture which can readily be seen indicates why it is believed by some that the expansion process may be slowing.

As can be observed from these numbers, it would be erroneous to say there exists a deceleration factor in the expansion of the Universe, for as stated elsewhere, it is strictly because of constant acceleration that, as time goes on, the recessional velocity between any two galactic clusters becomes less and less *per unit distance,* but then, this is simply due to the fact that each cluster continues to increase its velocity at a constant rate.

Look once again at the numbers in Fig. 1–1, The Universe In Numbers. Notice that after 9 billion years *galactic cluster No. 1* is moving through space at 49,248 miles per second and would naturally have a recessional velocity from time-zero of that same amount, 49,248 miles per second. Notice also that it is only about one billion of our present light years distant from time-zero. Now look at the recessional velocity between this same cluster No. 1 and between cluster No. 2 (located respectively at the one billion light year mark and the two billion light year mark). They too are one billion light years apart. However, even though galactic cluster No. 2 is traveling at a much higher rate of speed than cluster No. 1, their recessional velocity per unit distance of 1 billion light years has decreased dramatically to about 16,400 miles per second per billion light years, as in contrast to cluster No. 1 from time-zero of 49,248 miles per second. Now then, look at galactic clusters No. 6 and No. 7. Their velocities are constantly increasing at the rate of 5,472 miles per second per billion years, just as the cluster before them and after them do. Notice also, that they too are one billion light years apart, just as clusters No. 1 and No. 2, yet their recessional velocity is now only about half of what it is between cluster No. 1 and cluster No. 2 (about 8,000 miles per second per billion light years) much less than the 16,400 miles per second between cluster No. 1 and No. 2. And if we look at galactic cluster No. 17 (our own local group of galaxies) we will see that the recessional velocity between us and cluster No.16, or No.18, is now only about 5,472 miles per second per billion light years. At this point in space-time we have traveled for 34 billion years for a distance of 17 billion light years. I remind the reader once again however, that unless otherwise specified, a Light Year as applied in this sense always refers to our present *5.9 Trillion Mile Light Year* (usually referred to as 6 trillion for convenience) even though the real length of a light year as depicted by The Universe In Numbers depends upon the velocity in which the galaxies in their time frame move through space. For as is shown throughout this book *the velocity of light strictly derives from, and mimics the velocity of the emitting body.*

Once again, getting back to the "mechanics of velocities and recessional velocities of galaxies," let us look at galactic cluster No. 68. It will be seen that cluster 68 is twice as old as our local group which the Milky Way is part of. We see that cluster No. 68 is *68 billion years* old but is now four times the distance from time-zero that the Milky Way is, and is traveling twice as fast, 372,000 miles per second. Therefore, the velocity of light for any hypothetical inhabitants of that galaxy would likewise be about 372,000 miles per second!

Of greater significance however, *concerning deceleration,* is the fact that even though galactic cluster No. 68 is traveling through space at twice our present speed of 186,282 miles per second, the recessional velocity between it and galaxy No. 67 (*which is one billion of our light years away from 68*) is now only 2,700 miles per second—or 2.7 miles per second per million light years in that particular and far out vicinity of space.

These then may be the reasons why some astronomers suspect that the expansion of the Universe is slowing, *i.e., decelerating.* It seems only to appear that way because astrophysicists have not heretofore seriously considered that the Universe may actually be expanding under the influence of the same basic Newtonian laws of physics which apply to freely falling bodies here, for example, on Earth.

It would seem from all of this then that in order to *have order* and be able to solve some of the many problems which cosmology presently confronts us with, we may once again have to give to Newtonian mechanics the full front seat it once enjoyed.

Cutting the Hubble Constant in Half

In Fig. 1–1 of The Universe in Numbers there still remains one major discrepancy to account for, and that is the difference between the Universal Constant of Acceleration of about 5.5 miles per second per million years, *or per million light years in our immediate vicinity,* and the present estimated value of the Hubble Constant of 11 miles per second per million light years.

What The Universe in Numbers (Fig. 1–1) tells us is that images of galaxies reaching us at our present space-time position did not get here by propagating at one constant velocity of 186,282 miles per second, but rather, they arrived here at Earth at various velocities (see Fig. 1–1a)

The Universe in Numbers

An observer on a galaxy with an age of:	Is at distance from t=0 of:	Moves with that galaxy at a velocity of:	Can observe in direction of t=0, birth of another galaxy with age of:	Whose actual distance is half its own distance from t=0 or:	Moving at a velocity of:
12 BY's	2 BLY's	66/M mps	9 BY's	1 BLY's	49/M mps
17	4	93	12	2	66
21	6	115	15	3	82
24	8	131	17	4	93
26	10	142	19	5	104
29	12	159	21	6	115
31	14	170	22	7	120
34 (MW)	17	186	24	8	131
37	20	202	26	10	142
48	34	263	34	17	186
68	68	372	48	34	263

Fig. 1–1a. The Numbers In Fig. 1–1a above are approximate. They refer to The Universe In Numbers (Fig. 1–1) and reflect the greatest distance an observer on any galaxy can theoretically detect any material substance in the direction of t=o (the inner Universe) which is always younger than Observer's Galaxy. See for example our own position at 34 Billion Years in time, the Milky Way. This does not apply as such when an observer is looking towards the "outer edge" of his Universe. In that case, observer cannot observe any hypothetical births of galaxies, but can observe galaxies as they were when they were the same age and distance from t=o as observer's galaxy, *even though they actually have different ages,* (all older than observer's galaxy) and are physically at various distances from observer's galaxy.

as refers to the *Inner portion of our Universe,* the portion in the direction of t=0.

As refers to the outer portion of our Universe, we simply run into the faster moving galaxy's zero-velocity (0-c) images. Of course, not realizing this, we simply believe the images themselves had traveled to us with a velocity of 186,282 miles per second, *our own present speed through space,* when in fact it is we ourselves who have traveled and have run into the images of the galaxies which are older than our own galaxy.

Therefore, the manner in which we presently determine recessional velocities should be revised to allow for the fact that The Velocity Of Light Is *NOT* Absolutely Constant.

As a consequence, our distance scale would then have to be doubled in order that the calculation of the Hubble Constant (presently estimated to be about 57 kilometers per second per megaparsec) will reflect a more accurate and realistic recessional velocity of *exactly half the present estimate*, or about 28.5 kilometers per sec per megaparsec. This is about 5.5 miles per second per million light years, or, approximately 2.8 Angstroms per second per light second.

As can be seen from the numbers picture in Figures 1–1 and 1–1a, Hubble's Constant is exactly twice the value of the Universal Constant of Acceleration which I have here proposed. Also, the Hubble Constant applies to light years—a *distance* scale—whereas my own estimate of the constant applies strictly to years—a *time* scale.

For all practical purposes however, both constants can be reasonably associated with either distance or time, but only within about a *3 to 4 billion light year radius of the Milky Way* as previously pointed out, but not more than possibly 5 to 6 billion light years at the extreme. In the long run however, the Universal Constant of Acceleration, 2.8×10^{-10} m/s², must be based *strictly upon time, not distance, i.e., years, not light years*.

So how then are we to account for this disparity which exists between the Hubble Constant of Recession of 11 miles per second per million light years of *distance*, and the Universal Constant of Acceleration of 5.5 miles per second per million years of *time*? Remember, units of distance or time within about 5 billion light years of the Milky Way will make no substantial difference—it will only begin to matter out past 5 to 6 billion light years or more.

As I have pointed out previously, the Hubble Constant

of Recession, H_o, is exactly twice that of the Universal Constant of Acceleration, a_u. The only real defense then of my own constant—which is approximately one-half the value of a recent estimate of H_o as put forth by Sandage* —is simply that the *average velocity of Light must be reconsidered to be 93,140 miles per second when estimating its time in getting to our telescopes* (instead of the 186,282 miles per second) since the most significant and most important statement of this book is that the precise velocity of light is "Not" after all constant in the most technical sense of the word.

On this basis then, when astronomers declare a recessional velocity between us and any other galactic cluster, (based upon a Hubble Constant of 11 miles per second per million light years,) the 11 miles per second should then, *in reality,* be 11 miles per second per *two (2) million light years,* since it is the recessional velocity which is actually determined; and from that, the distance is calculated accordingly. And since the velocity is determined on the basis of light's velocity always having been steady at 186,282 miles per second, the Hubble Constant was then determined to be 11 miles per second per million light years. However, if we now consider that light's velocity has constantly increased, then the whole basis for determining distance must be revised to reflect light's average travel time to be slower in reaching us by a factor of one-half. Therefore, what we had believed to be one million light years away from us, is in reality, two (2) million light years away. So that the Hubble Constant should now reflect about 11 miles per second per two (2) million light years instead of one (1) million light years. That of course brings it around very nicely then to the within stated estimate of 5.472 miles per second per million light years, which in "Angstrom Units" is 2.791 Angstroms per second per light second.

*Telephone conversation with Allan Sandage.

9

On the Way Light Propagates

During the late 1880's, Albert A. Michelson and Edward W. Morley were having a difficult time attempting to prove that light propagated in strict accordance with the "Principal of the Addition of Velocities," such as is the case, for example, with acoustical radiation.

Because of the Earth's orbital velocity around the Sun, Michelson and Morley attempted to prove that they could detect a difference in the velocity of light by directing two beams of light at a 90 degree angle from each other. And since the Earth orbits the sun at about 18-mps, they had naturally thought this difference in light's velocity should show up when they compared one beam's velocity from the other. They were, however, never able to prove a variation existed.

Sixteen years later, in 1905, Albert Einstein, in his Special Theory of Relativity, declared that the speed of light was an *Absolute and Unvarying Constant throughout all space and time*—even to the extent that if one could move alongside a light-beam at 90% it's speed, they would still measure the beam to be moving at 186,000 mps.

The sketches in this chapter outline what I believe is more properly happening when light propagates; and why Michelson and Morely were not able to measure the difference in variation of light's velocity. In other words, beside the fact that light constantly increases by 2.8 Angstroms per second per second, it also varies in its propagating speed, from zero velocity (0-c), to 372,000 miles per second (2-c), as figures (1–2) to (1–12) will purport to show. What follows then will be a sketch of each figure, and a brief description. Note that the "Earth and Moon," *for all practical purposes,* are depicted as being exactly one (1) light-second (186,000 miles) apart, even though they are, in reality, about 240,000 miles apart.

In Figure 1–2, a theoretical observer on another galaxy located at Milky Way Image position 2-c, hypothetically sees Milky Way's image of its birth. At that moment, Milky Way's birth image has just arrived at position 2-c as observer galaxy (located 34 BLY's from t=0) is just "at that moment in time" moving away from 2-c image of Milky Way at 265,000 miles per second. Therefore, even though in reality, the actual Milky Way itself is located at 1-c (17 BLY'S from t=0) traveling at 186,282 mps, its birth image is hypothetically observed by observer at 2-c position as having a recessional velocity between them of 265,000 mps, and is perceived to be at a distance from observer galaxy of twice its real distance, since observer at 2-c is only seeing *at that moment in space-time* the Milky Way as it was at birth, that is, at its t=0 (0-c) space-time position. So in effect, when observer at 2-c sees the image of the Milky Way as it was at birth, the Milky Way itself is actually only half that distance away in a much more developed stage which observer at 2-c cannot yet see.

According to Newtons Law of Motion, the amount of

On the Way Light Propagates

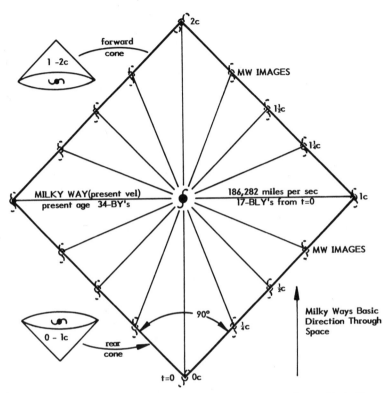

Fig. 1–2. The Milky Way's Two Light Cones—From (0-c) Zero Miles Per Second to (2-c) 372,564 Miles Per Second.

velocity which can be imparted to a body will depend upon the body's mass and the force applied to it; and that any velocity once imparted will be rigidly maintained unless other forces act upon it, i.e., when no force is exerted on a body it remains at rest or continues on in a straight path at constant speed, whichever is the case. Naturally then, the more force that is applied to a body, the faster it will move; and of course, the less massive a body is, the faster that same force will impel it.

In empty space then, where there is normally presumed to be no frictional effects acting upon a body, an object set in motion would simply continue in motion in a straight line with constant speed unless some other force acted to change its direction and or speed. How fast the object travels will depend strictly upon its mass and the amount of energy imparted to it. This is known as the Law of Inertia, *Newtons first law of motion,* which had first been proposed by Galileo before him.

Let us go on then to a *Photon of Light* and apply this Principle of the law of motion to a light beam. Light beams are made up of photons, and photons are presumed to be massless. Therefore, if photons have no mass, and the source from which the photons issue is already in motion—such source moving at a speed of 186,282 mps—what will be the photon's velocity? The answer to this should be evident, for if photons are massless they should propagate away from the source in its forward direction of travel at the precise speed of the moving source which they (the photons) are released from. So that relative to space, the photons will actually travel a total distance of 372,564 miles in one second. In other words, the photons are being impelled out ahead of the source at 1-v (186,282 mps) while at the same time they are being carried along with the source at another 1-v for a total velocity (relative to space) of 372,564 mps, or 2-v (see Fig. 1–3, Propagation of Light at 2-c). This is in strict accordance with the Principle of the Addition of Velocities. However, it will appear to all observers (who themselves do not realize it is the source itself and they who are actually moving at 1-v) that the photons *on their own* traveled 186,282 mps, or 1-c.

If released from the rear of a moving platform (of which platform we will refer to here as the Earth) the photons will simply stay put, *dropping off in the Earth's tracks,* just as do the *contrails of hot gases* from jet planes in flight. In this case however, even though the actual velocity of the photons (aimed, or otherwise pointed in the rearward direction) is zero, 0-c, it will still appear to all observers that the photons had shot away from the Earth at 186,282 mps, 1-c, since the observers themselves are actually moving at that speed along with the Earth, but in fact do not realize that they (the photons) are not moving at all. (see fig. 1–4, Propagation of Light at 0-c).

If the photons from the light source were released from the side of the moving platform, that is, directed at a 90 degree angle to the platform's forward motion (see fig. 1–5 Propagation of Light at 1-c) they would then propagate out at the precise velocity of the body they are released from—186,282 mps—a speed of 1-v, or once again, 1-c, as refers to the speed of light.

This then is why light always appears to propagate at one speed no matter how it is measured. It is because we have not been aware of the most significant factor, and that is that it is we (the Earth, the stars, the galaxy) that are actually moving through space at the velocity we attribute to light. And so, no matter how, or from which direction we measure the velocity of light, it will always appear to us to propagate at the one unvarying speed of 186,282 miles per second—at least for this moment in space-time—since for each second of time which goes by, our own basic velocity actually increases by approximately 2.8 Angstroms; consequently, so too will the velocity of light follow suit and increase likewise.

This, most apparently then, is the sole reason why Michelson and Morley were not able to detect a variation in the speed of light relative to the Earth's velocity in its orbit around the Sun. It was simply because the propagation of light itself exists solely as a result of the fact that the body in which it is emitted from, is itself in motion moving through space, and that it is this precise velocity of the emitting body which determines the precise velocity of light as we perceive it to be. Therefore, since the propagation of light strictly depends upon the real and actual velocity of the body it is emitted from, it then stands to reason that light's velocity, *relative to space,* should in reality, vary in every conceivable direction—from as much as twice (2-c) the emitting body's real velocity, all the way down the scale to zero-velocity, 0-c.

However, as was the case with Michelson and Morley's experiments, there is no known way we can differentiate between the varying velocities of light, since relative to the observer, light will always appear to have but one velocity (that is the velocity in which the observer moves through space) no matter which direction the observer looks in, or how it is measured. For example, Figure 1–2 shows a schematic of the propagation of light from our own galaxy, the Milky Way. This can represent any galaxy, star, planet, or for that matter any mode of electromagnetic radiation located anywhere in the Universe. It is designed to depict only the general picture of how light propagates at varying speeds (relative to space) from zero-velocity, 0-c, to twice the velocity, 2-c, of the emitting body's real velocity, 1-v.

Starting with the Milky Way's image, both in the direction of 0-c, and at 0-c itself, what happens here is that as the Milky Way moves ahead, its image just simply stays in place, *drops off from the rear of galaxy,* just as the contrails of hot gases issue, or drop off from the rears of jet engines of planes in flight. In that direction, the actual velocity of light "relative to space" is zero, 0-c. However, since all observers on our galaxy are moving along with the galaxy at 186,282 miles per second (but do not in fact realize it) they will naturally determine the velocity of any light beam they point in the 0-c direction to be 186,282 mps (see Figure 1–4 Earth-Moon Schematic).

Going now strictly to the Milky Way's 2-c direction of its image in Figure 1–2, what happens here is that the image is propelled out ahead due to the Milky Ways forward momentum. It is propelled at the precise speed the Milky Way is traveling (Newton's 2nd Law of Motion) so that whatever distance the Milky Way has actually traveled from t=0, its earliest *at birth image* has also traveled out ahead of it an additional and equal distance, so that at that point in space the velocity of light, *relative to t=0,* is 372,564 miles per second, 2-c. Observers however, on another galaxy located at 2-c would measure their recessional velocity with the Milky Ways birth image to be 265,000 miles per second, since at that space-time position, an observer at 2-c is actually traveling at 265,000 mps. This observer will measure his velocity of light to be 265,000 mps for the same reason we measure ours to be 186,282 mps. In other words, the velocity of light is never really constant, but is strictly related to a body's space-time position, i.e., the body's real velocity in space from time-zero, t=0. Likewise, since observers are moving along with the Milky Way at 186,282 mps, they will naturally measure their velocity of light to be 186,282 mps no matter how they measure it. See for example, Figure 1–3, Earth-Moon Schematic.

What happens also at galaxy image (1-c) in fig. 1–2 is simply that the Milky Ways forward momentum causes its image to be propelled out at right angles at the same speed that it travels. Because the galaxy moves forward it drags its image right along with it, and although the image appears to travel the long path of the *hypotenuse,* it actually travels the shorter distance of the lines at right angles, hence, the velocity of light in this case is precisely 186,282 mps, 1-c, both for an observer, *and relative to space.* Further details of this can be seen in Figure 1–5, Earth-Moon Schematic. In addition, Figures 1–6 and 1–7 detail the light rays at velocities of ½-c and ¼-c respectively.

In Figure 1–3 the Moon and Earth are moving together through space, the Earth directly behind the Moon at a velocity of 186,282 mps. For the sake of convenience they are positioned precisely one light second (186,282 miles) apart from each other instead of their actual average distance of 234,000 miles.

How does the laser beam from Earth arrive at the

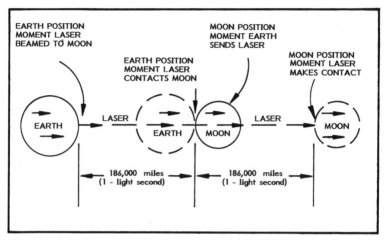

Fig. 1–3. Propagation of Light at (2-c) 372,564 Miles Per Second.

Moon? What is the basic mechanism of propagation? It is simply that as the Earth moves forward at 186,282 mps, the beam, having no mass to speak of, is automatically propelled out ahead of the Earth at precisely the same speed in which the Earth moves through space (this in accordance with Newton's laws of motion). By the time 1-second has passed, when the Earth has advanced 186,282 miles to the position the Moon was originally at when the beam was initially directed to the Moon, the Moon will also have advanced 186,282 miles.

What actually happens then, is that the Earth carries, or pushes the beam along with it at the velocity it is traveling, 186,282 mps; and in addition, it also propelled the beam out ahead of it at another 186,282 mps (according to Newton's second law of motion) so that in effect, the beam traveled—relative to the original position of Earth when it released the beam—372,564 miles.

In Figure 1–4, both Moon and Earth are moving together through space, the Moon directly behind the Earth at a velocity of 186,282 mps. For the sake of convenience

On the Way Light Propagates 71

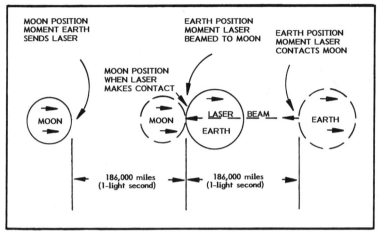

Fig. 1–4. Propagation of Light at Zero Velocity (0-c).

they are positioned precisely 1-light second (186,282 miles) apart from each other instead of their actual average distance of 234,000 miles.

How does the laser beam get to the Moon from the Earth? What is the actual mechanism of propagation? It is simply that as the Earth moves in its forward thrust through space, the beam that is aimed at the Moon from the Earth's rear position just simply stays put, i.e., it drops off in its tracks as the Earth pulls away from the beam—once again—just as the contrails of hot gases drop off in the tracks of jet planes in flight. So that in effect, the Moon runs head on into the laser beam at the speed with which the Moon moves through space, 186,282 mps.

Here again however, observers on Moon and Earth do not realize it is they that are moving through space at 186,282 mps, and therefore, both simply believe the laser beam—on its own—propagated from Earth to Moon at 186,282 mps, when in reality the beam actually traversed space at a speed of 372,564 mps since it was pushed, or otherwise carried along with the Earth for 186,282 miles in that one-second journey, while at the same moment, it was thrown out or otherwise propelled ahead of the Earth for an additional distance of 186,282 miles.

In Figure 1–5, the Moon and Earth are moving together through space (side by side) at a velocity of 186,282 mps. For the sake of convenience they are positioned precisely 1-light second (186,282 miles) apart from each other. How does the laser beam from the Earth arrive at the Moon? Not so simple! As the Earth and Moon move forward through space, the beam from Earth is aimed at the Moon from a 90 degree angle in relation to Earth's forward thrust through space. As the Earth advances through space along with the Moon, side by side, each at 186,282 mps, the beam from Earth is naturally carried along with it. However, since it is directed from a 90 degree angle (in relation to its forward thrust through space) towards the Moon, the speeding Earth also hurls, or shoots the beam out at precisely the same velocity the Earth moves (this again according to Newton's second law of motion). The Earth's forward thrust through space of 186,282 mps is precisely applied to the massless laser beam—aimed from a right angle of its forward thrust direction—so that there is a misleading appearance of the beam actually traveling the longer length, *geometrically known as the hypotenuse in this triangular sketch,* when in fact it is continuously propagated from Earth at right angles—relative to their forward motion through space—for a total distance of 186,282 miles. Here again, observers on both Earth and Moon simply do not realize it is they who are actually moving through space at 186,282 mps; and so, they naturally believe the beam propagated in one single straight line at 186,282 mps, instead of the more broad and complicated path it took in reality.

In Figure 1–6, the Moon is one-light second away at a 45 degree angle behind Earth in their forward thrust through space. Since Earth releases laser beam at a 45

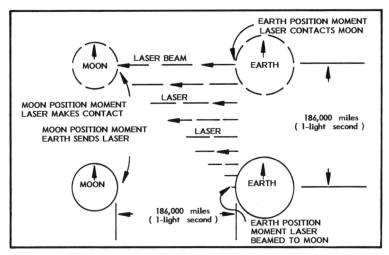

Fig. 1–5. Propagation of Light at (1-c) 186,282 Miles Per Second.

degree angle from its rear quarter position, the beams actual velocity relative to its space vector will thereby be half the speed the Earth itself travels, or 93,141 mps. The other half of the beams apparent velocity of 186,282 mps is due to the effect of the Earth pulling away from it. So then, as the beam is thrust towards the Moon, spreading at ½ the speed of Earth's velocity through space, it is also deposited in place as the Earth races away from it and the Moon rushes towards it; so that it will then appear to observers on Moon and Earth that the beam simply shot out from the Earth at 186,282 mps in one single line of propagation. Neither realize they have themselves each advanced through space 186,282 mps while the beam (in one-second of time) was finding its way to the Moon at only half the actual velocity which they themselves are moving—half the apparent velocity of light. It is, in other words, the propagation of light as a combination of the action which figures 1–4 and 1–5 depict.

In Figure 1–7, for the sake of convenience, Moon is

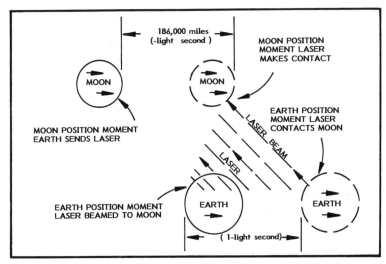

Fig. 1–6. Propagation of Light at (½-c) 93,141 Miles Per Second.

one-light second away, but only at a 22½ degree angle behind the Earth in their basic thrust through space. Since Earth releases laser beam at a 22½ degree angle from its rear quarter position, the beam's actual velocity—relative to space—will thereby be only one-quarter of the speed the Earth itself travels, or 46,570 mps. The other ¾ of the beam's apparent velocity of 186,282 mps is due to its staying put, *just simply dropping off in its tracks and spreading,* as the Earth moves away from beam, and the Moon rushes towards it. So that as the beam spreads at ¼ the speed of the Earth's velocity through space, it is also deposited in place as the Earth pulls away from it. Consequently, it will then appear to observers on Moon and Earth that the beam simply propagated from Earth at 186,282 mps in one single line of travel.

They will not realize they have each advanced through space 186,282 miles as the beam, in one second of time, was finding its way to the Moon at only ¼ the actual velocity

On the Way Light Propagates 75

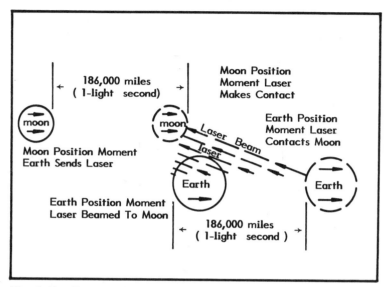

Fig. 1–7. Propagation of Light at (¼ c) 46,570 Miles Per Second.

which they themselves are moving, or, ¼ the apparent velocity of light.

The schematics in Fig. 1–8 depicts the ages and distances traversed by three galaxies, X, Y, Z, and their *double light cones*. The distance scale is based upon our own present light year of 5.9 trillion miles. The time scale is based upon our own length of year of 31.6 million seconds.

Notice that the distance in light years is not equal to the ages in years. That is simply because they are based upon our own space time standards in this schematic. Also, it is no coincidence that galaxy Y, the Milky Way, is 17 billion light years in distance from time-zero and exactly twice that number, 34 billion years in age. As stated previously, this 2-to-1 ratio is not unique to us. Again, that is simply because our present light year scale is utilized in the calculations (See for example Fig. 1–1, The Universe In Numbers).

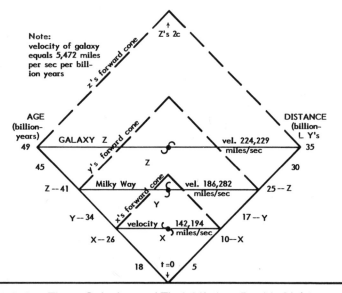

Fig. 1–8. Three Galaxies and Their Lifetime Double Light cones.

The three double light cones in Fig. 1–8 are due to the fact that a galaxy's light—in its forward direction through space—is propagated at velocities between one and two times the actual velocity of the galaxy from which the light issues. For example, galaxy Z's image propagated at twice its velocity, 2-c, or 448,704 mps as indicated at the extreme upper point of its broken line cone, and has propagated at one-time its velocity, 1-c, or 224,352 mps to the extreme right and left of its actual position. However, at the extreme bottom portion of its solid line (rearward) cone, its velocity of light is zero (0-c).

The velocities of light in between these examples can simply be determined by the process of extrapolation. Any light source, whether its a galaxy, a star, or a planet, or for that matter any other emitting source of electromagnetic radiation, will naturally create a double cone of its signal, image, etc. and will propagate through space at velocities ranging between zero (0-c) and twice (2-c) the actual velocity of the source itself. However, the apparent velocity of 1-c (as in contrast to the real velocities 0-c to 2-c) will always seem to be at 1-c no matter which direction it is measured from, and no matter how it is measured, just as the famous Michelson-Morley experiment proved to be when they themselves attempted to detect the variations in the velocity of light. For additional clarification on the variation of light's velocity see Figures 1–3 through 1–7.

In Figure 1–9, galaxy A represents the Milky Way at a distance of 17 billion light years from time-zero after 34 billion years of constant acceleration at 2.8×10^{-10} meter per second per second (approximately 2.8 Angstroms per second per second). Its image (light) is propagated in all directions—from zero velocity, 0-c, to twice its actual velocity, 2-c. It is this variation of the velocity of light (from zero to twice that of the actual velocity of the emitting body) which creates the double light cone. This same picture applies to galaxies B, C, and D. The only differences are that B is 14 billion of our present light years from time-zero after 31 billion years of travel at a constant acceleration of 2.8 Angstroms per second per second, while C is 11 billion light years from time-zero after 27 billion light years of travel at the Universal Constant of Acceleration as in B above; and D likewise, is 9 billion light years from time-zero after 25 billion years of travel.

Their present velocities through space appear in the schematic of Figure 1–9. It is not simply a coincidence that our own galaxy's age of 34 billion years is exactly double in "years of time" as its 17 billion "light years of distance." As can be seen from Figure 1–1 (The Universe In Numbers) this 2-to-1 ratio is no exception—but is rather the rule for all galaxies throughout the Universe. This is simply because different velocities will naturally result in different light year distance scales; and so, based upon any galaxy's present light year, that particular galaxy will

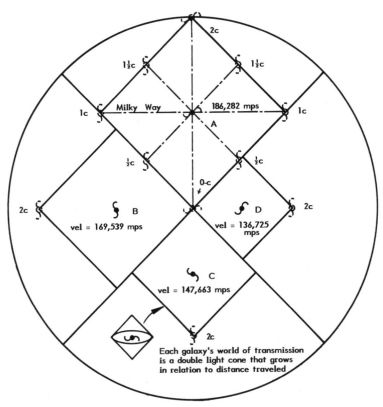

Fig. 1–9. The Universe—Its Galaxies And Their Light Cones.

always be twice its age in years that it is in distance of its own light year from time-zero.

In Figure 1–10, each and every material body in the Universe, whether a particle or a star transmits its image, signal, etc. in the form of a double (face to face) cone. These double light cones are the result of the body's basic forward thrust through space. The distances they will extend to will be equal to the actual distance the body traveled. Naturally though, a body's temperature, and or signal strength, will always affect the extent of its observable trans-

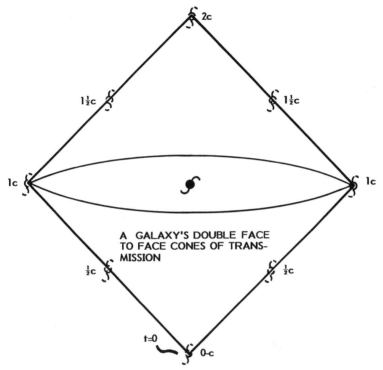

Fig. 1–10. The Double Face To Face Cones of Transmission.

missions, as for example, with very bright Quasars—for although their light cones naturally exist by virtue of their travel time from t=0, their extremely bright light deems them to be more readily detectable.

In Figure 1–11, Uniform Expansion can have but one interpretation—and that is—Acceleration! Otherwise, it would be impossible to say that the Universe is expanding *in the Hubble sense* if we do not mean to imply that the galaxies are *accelerating* at some constant rate.

The picture of 1–11 is no different for galaxies flying apart from each other, i.e. *accelerating,* than it would be for a quantity of golf balls falling inwardly towards each other to Earth under the influence of Earth's gravitational

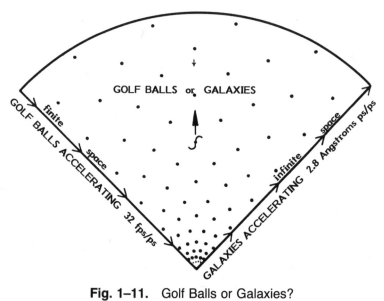

Fig. 1–11. Golf Balls or Galaxies?

field. However, the golf balls would initially start out from the top of the picture in Figure 1–11 and accelerate towards the center of the Earth at a constant rate of 32 ft. per second per second.

The galaxies motions differ from the golf balls in that the galaxies accelerate in the opposite direction towards the top of the cone in a "spreading fashion" at only a minuscule rate of acceleration of one-ninety-millionth of an inch (2.8 Angstroms) per second per second. Otherwise, the laws of nature for both systems are identically the same, that is, the doubling of time at any constant rate of acceleration always equals a quadrupling of distance traveled. And that is so, simply because in any system of constant and uniform acceleration, a doubling of time means also, a doubling of the velocity. That of course is why distance quadruples at twice the time.

Figure 1–12 shows a sample of the Universe in which

On the Way Light Propagates

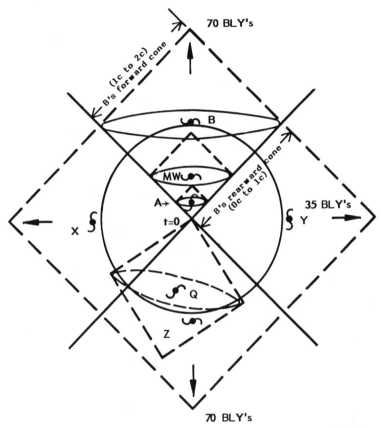

Fig. 1-12. Who Can Observe Who In This Slice Of The Universe?

MW represents the Milky Way moving at its present velocity of 186,282 mps. MW is 17 billion light years from time-zero after accelerating through space for 34 billion years at a constant rate of 2.8×10^{-10} meter per second per second. Its forward light cone extends an additional 17 billion light years so that its image at birth is detectable at 34 billion light years from time-zero at a point in space just short of galaxy B. The outer edge of this sampling of the material universe (galaxies B, X, Y, and Z) extend out to 35 billion light years from t=0, while each of their forward light cones extend out to 70 billion light years from time-zero.

In this scenario, galaxy A can hypothetically observe galaxies MW and B providing that its observers possess sufficient detection power. MW can observe B but not A since A's forward light cone does not extend to MW. B cannot yet observe MW's birth since MW's forward light cone comes just short of reaching B; and of course, B cannot observe A. Z can observe Q, and Q can observe Z. Neither X nor Q receive each others light, for as Figure 1–12 depicts, although each of their light cones extend into the others light cone, neither light cone extends out to either galaxy.

10

How Mass Is Created

The process of how protons, neutrons, electrons, etc. acquire mass, has remained a mystery. However, in view of the *Outside Force* which appears to be *Eternally Accelerating the Particles* in the Universe at a constant rate of 2.8×10^{-10} m/s^2, the precise amount of mass a particle acquires should therefore be strictly determined by the ratio of its rate of acceleration to its length, of which I will refer to here as *Particle Acceleration Rate (PAR), in units of the particle's diameter.* And since Einstein's energy equation, $E=mc^2$, teaches that mass and energy are interchangeable, then any particle which accelerates at this constant rate should naturally acquire a quantity of mass-energy in direct relation to its *length,* so that the smaller a particle is, the more massive it will be.

A particle's *Rest Mass* might then be determined simply by dividing it's length into the Universal Constant of Acceleration, a_u, which will result in a quantity describing both, the Particle Acceleration Rate, PAR, in units of its diameter, d; and, equivalently, its mass, m, strictly in units of "keVs" (thousand electron volts); both of which turn out to be one and the same quantity. And since a naturally accelerating particle, as I have herein propounded, creates mass and energy in strict relation to its *length*, it should readily be seen why Mass and Energy are interchangeable; and naturally, why the *smaller* particles which make up the atom possess more mass and energy than their *larger* counterparts.

To determine then, the diameter of the proton, d_p, when we know either its mass in units of keVs, or it's particle acceleration rate per second, we would only need to divide its particle acceleration rate, PAR_p into the Universal Constant of Acceleration, a_u, so that,

$$d_p = a_u/PAR_p \qquad (5)$$
$$d_p = 2.791 \times 10^{-10} \text{m/s}^2 / 9.382 \times 10^5 \text{ per sec}^2$$
$$d_p = 2.974 \times 10^{-16} \text{m}$$

As has been stated then, the proton's mass (in keV units) appears to be equal in quantity to its natural rate of acceleration in *Units of its own Length;* so that mass and energy appear to be strictly dependent upon *particle length;* and are therefore determined simply as the ratio of its length to the Universal Constant of Acceleration, a_u. In this manner, for example, a proton's constantly manufactured energy (its particle acceleration rate, PAR), is converted to mass, just as $E=mc^2$ predicts. So that we will then have the following equation for determining Particle Acceleration Rate, PAR (rest mass) of a proton such that

$$PAR_p = a_u/d_p \qquad (6)$$
$$PAR_p = 2.790\ 625\ 94 \times 10^{-10} \text{ m/s}^2 / 2.974\ 217\ 516\ 8 \times 10^{-16} \text{m}$$
$$PAR_p = 9.382\ 723\ 1 \times 10^5 \text{ per sec}^2$$

How Mass Is Created

And if a_u were not known, but particle length and particle acceleration rate were known, we could then of course identify the Universal Constant of Acceleration to be the product of "length and mass" of any particle in question; i.e., knowing a proton's diameter, d_p, and its particle acceleration rate, PAR_p (a quantity whose quotient appears to be precisely equal to its mass in keVs) we can then determine for a_u

$$a_u = d_p \, PAR_p \qquad (7)$$
$$a_u = (2.974 \times 10^{-16} \text{m})(9.382 \times 10^5 \text{ per sec}^2)$$
$$a_u = 2.791 \times 10^{-10} \text{ m/s}^2$$

The *King of all the particles* may most likely be the *Electron*, where I have speculated that Planck's Constant, h, *that indivisible packet of energy,* whose value in electron-volt seconds is 4.135 669 2×10^{-18} keV-second, may *in itself* actually be the direct result of the electron's acceleration rate, PAR_e, by one length, (one-diameter of 4.135 669 2×10^{-18} meter) due to its natural acceleration, a_u. I suspect here that this would most likely be the electron's *VIRTUAL* point-like diameter and should be equal to the same quantity in "units of meters" as Planck's Constant is in "units of keV-seconds"—once again indicating why mass and energy are interchangeable. Here however, I have adjusted the numbers only very slightly (a_u was originally 2.788×10^{-10} m/s^2) because this is where it appears to me to be that Planck's Constant, h, must originate from—for when we divide h into the product of a_u and a proportionality constant, k, which is designated to be equal to 1-m/sec, we come out with a quotient for the electron's mass, *in units of keVs*, so that

$$m_e = a_u k / h \qquad (8)$$
$$m_e = (2.791 \times 10^{-10} \text{ m/s}^2)(1.0 \times 10^0 \text{ m/s})/(4.136 \times 10^{-18} \text{keV-sec})$$
$$m_e = 6.748 \times 10^7 \text{ keV's}$$

The *Electron's Virtual Mass* appears then to be about *72-times heavier* than the mass of the proton; however, it would not normally be measured as such. Its particle acceleration rate, PAR_e would therefore be determined by dividing a_u by its point-like diameter of $4.135\ 669\ 2 \times 10^{-18}$ meter (the same quotient as h, except the unit changes from one of "Energy" in keV-sec, to one of Length) such that

$$PAR_e = a_u/d_e \qquad (9)$$
$$PAR_e = 2.791 \times 10^{-10} \text{m/s}^2 / 4.136 \times 10^{-18} \text{ meter}$$
$$PAR_e = 6.748 \times 10^7 \text{ per sec}^2$$

and in order to determine both, Planck's Constant, h, and the *virtual point-like* diameter of the electron, d_e (of which, I repeat, both appear to be one and the same quantity) we would simply divide the Universal Constant of Acceleration, a_u, by the electron's particle acceleration rate, PAR_e, to arrive at

$$d_e = a_u/PAR_e \qquad (10)$$
$$d_e = 2.791 \times 10^{-10} \text{m/s}^2 / 6.748 \times 10^7 \text{ per sec}^2$$
$$d_e = 4.136 \times 10^{-18} \text{m}$$

Notice that even though the unit above results in meters, this quantity is the same for "h" in units of keV-sec, possibly indicating where "h" derives from

To arrive at the same value of Planck's $h/2\pi$ utilizing a_u, we need only convert the mass in energy units of the electron, to units of mass in kilograms. So that $h/2\pi$ will be equivalent to the product of the electron's mass in kilograms; the Universal Constant of Acceleration, a_u; (π); and the proportionality constant, k, designated as 1.0×10^0 m/s, so that

$$h/2\pi = m_e a_u \, \pi \, k \tag{11}$$
$$h/2\pi = (1.203 \times 10^{-25} \text{kg})(2.791 \times 10^{-10} \text{m/s}^2)(3.141 \times 10^0)$$
$$(1.0 \times 10^0 \text{m/s})$$
$$h/2\pi = 1.054 \times 10^{-34} \text{ Joule-Sec}$$

and to determine Planck's Constant h in association with the Universal Constant of Acceleration, we would retain the above equation except that (π) must be replaced with $2(\pi^2)$, such that (h) in units of joule-seconds can be arrived at by

$$h = m_e a_u \, 2(\pi^2) k \tag{12}$$
$$h = (1.203 \times 10^{-25} \text{kg})(2.791 \times 10^{-10} \text{m/s}^2)(1.974 \times 10^1)$$
$$(1.0 \times 10^0 \text{m/s})$$
$$h = 6.627 \times 10^{-34} \text{Joule-Sec}$$

The above equations would tend to indicate that the quantum of action (Planck's Constant h) is precisely the same quantity as the electron's "VIRTUAL" point-like diameter is in units of meters. Therefore, the actual mechanism which reflects and determines the value of Planck's Constant, may, in retrospect, appear to be none other than the specific energy of the electron's natural rate of acceleration, a_u, in which it eternally increases its velocity at a rate of *one-electron diameter* of 4.135×10^{-18} meter for each period of elapsed time of approximately 1.5×10^{-8} second. In other words, the electron "VIRTUALLY" increases its velocity (due to its natural rate of acceleration, a_u), at a rate of approximately 10^{-18} meter per 10^{-8} sec per 10^{-8} sec.

This would mean that electrons appear to increase their *measurable mass* just as protons do. However, wherein the proton's energetic and constant increase in velocity is continuously and actually *added* to its rest mass, the electron's manufactured energy of eternal acceleration is *continuously released as energy of electromagnetic interactions,* and therefore, should not add nor subtract from its classical mass.

It would seem to appear then, that we are dealing here with two forms of energy: the *proton's energy* of constant acceleration, a_u, which eternally *increases its actual mass;* and the *electron's energy* of constant acceleration, a_u, which instead, is *converted to the various forms of electromagnetic phenomena.*

In this, it would seem that the constantly increasing energy due to a_u, (as manufactured over an elapsed time of one-second) should continuously increase the proton's actual mass in accordance with the following equation:

$$\Delta m_p = m_p (a_u/1.0 \text{ sec})^2 /c^2 \qquad (13)$$
$$\Delta m_p = (1.672 \times 10^{-27} \text{kg})(7.788 \times 10^{-20} \text{m/s})/(8.988 \times 10^{16} \text{m/s})$$
$$\Delta m_p = 1.449 \times 10^{-63} \text{kg/sec}$$
or—
$$\Delta m_p = (1.449 \times 10^{-63} \text{ kg/sec})(5.609 \times 10^{32} \text{keV/sec})$$
$$\Delta m_p = 8.127 \times 10^{-31} \text{keV/sec}$$

Of course, any increase in mass should have an effect on a particle's length, and should in this case cause the proton's diameter to correspondingly *contract* with time so that its diameter could well be approaching the 10^{-16} meter I have previously estimated it to be.

As for the electron, its manufactured energy due to its acceleration for each second should amount to:

$$\Delta m_e = m_e (a_u / 1.0 \text{ sec})^2 / c^2 \qquad (14)$$
$$\Delta m_e = (1.203 \times 10^{-25} \text{kg})(7.788 \times 10^{-20} \text{m/s} / 8.988 \times 10^{-16} \text{m/s})$$
$$\Delta m_e = 1.042 \times 10^{-61} \text{kg/sec}$$
or—
$$\Delta m_e = (1.042 \times 10^{-61} \text{kg/sec})(5.609 \times 10^{32} \text{keV/kg})$$
$$\Delta m_e = 5.844 \times 10^{-29} \text{keV/sec}$$

As stated above however, electrons should not in themselves increase in real mass as do protons, which is likely explained by the fact that electrons move in *Orbiting Trajectories,* and therefore *radiate away* their eternally manufactured energy. And of course, this in itself should answer the question as to why electrons do not fall into their parent "protons" as they radiate away their energy. It is simply because they are constantly developing new energy due to a_u. In this manner then, they should eternally maintain sufficient energy to remain in orbit—each electron orbiting "each individual proton" in stable atoms. This, of course, would naturally "deny" the existence of Quarks and Empty Atoms . . .

As for why I prefer electrons orbiting each individual proton instead of orbiting the nucleus as a whole—it is extremely difficult to embrace a picture of the atom's structure which dictates, for example: that if all the *supposedly* empty space in all the atoms were hypothetically subtracted from the total volume of, say *one-million tons of steel,* the amount of solid material left over would not suffice to occupy the space which *one (1) grain of rice* would occupy in the normal everyday sense.

Another picture depicting this supposed emptiness of the atom would be somewhat analogous to *Peas* (the electrons) circling a *Basketball* (the nucleus) 10-to-20 miles away. This would be equivalent to the *Basketball* being situated at the *White House* in Washington, D.C., with *Peas* circling all the way out to the *Beltway.*

And what would the Earth itself boil down to if it were hypothetically voided of all the empty space which each atom is supposed to consist of? In this case, there would be so little solid material left, *after hypothetical compaction to their theoretical maximum limits,* that if only the solid stuff of the Earths atoms (protons, neutrons, electrons) were then formed into a cube, it would fit neatly upon *one-half of a football field!* That's right, the Earth would consist of no more than a *50-meter cube of solid matter!*

So, whatever the details of the atom's structure may actually be, it is inconceivable that all matter could actually be comprised of approximately 10^{15} parts *empty space* to one (1) part solid matter.

The above examples then, are striking representations of the picture one would have to accept in order to hold to a belief that *All* the electrons *NORMALLY orbit outside the nucleus.*

Why then all this so-called empty space in the atom? The answer to that question may simply lie in the fact that we did not realize that the atom may indeed be *eternally accelerating* at the constant rate of 10^{-10} meter per second per second (Bohr's Hydrogen atom in its ground state is 10^{-10} meter in length). And since we did not realize the atom may indeed be accelerating at the rate of approximately *one-atom diameter's length per second per second*, we therefore could not have taken this factor into consideration; but instead took it to mean that its *elusively increasing velocity* was part of its actual diameter instead of its natural and eternal rate of acceleration, a_u.

So then, instead of having atoms which are full of empty space (wherein the key factor contributing to that picture are the electrons normally orbiting way outside the nucleus, i.e., up to 100,000 times the distance of a nucleus diameter) it would seem to be much more realistic to believe that "each electron" normally orbits an individual proton, *as is the basic case with hydrogen atoms,* while neutrons play the major role of stabilizing the atom by keeping protons separated from protons by about one-fermi (10^{-15} meter) in order that their super-strong *Repulsive* force (which they, the positively charged protons, would exert upon each other) will not have the devastating effects which could be expected of such a force. In other words, without the neutrons normally keeping protons (in each nucleus) separated by their 10^{-15} to 10^{-16} meter diameter, the repulsive force which protons have upon each other would likely *"blow the atom apart."* However, it is most likely *this repulsive force of protons* which is the key in determining the density of matter.

As for this Atomic Model of Electrons orbiting individual protons, the electrons *of course* would still be found outside the nucleus, as is commonly the case; however, this would be due to those electrons which are orbiting the atom's outer protons; and there are most likely as many, or more of these situations, than there would likely be of electrons orbiting inner protons.

This "Not So Empty Atom" would preferably result in the structure of atoms being more like that which is represented in Figure 1–13a. So that instead of, say, the helium atom being constructed as in figure (b), it would be constructed more like that of (a). And of course, this type of "atomic construction" would be representative of all atoms, no matter how many protons and neutrons they consist of.

In the schematic of Figure 1–13a, protons would not normally be next to each other, which in essence should cause the atom to be unstable and, consequently, annihilate. And of course, electrons would have open space to orbit freely around each inner or outer proton, *there being four different paths each electron would be free to take.* Also, the nucleus, still being approximately 10^{-15} meter in diameter, would, in essence, *be the complete atom;* so that the total empty space in an atom should then amount to not more than possibly *two parts space to one part solid, or possibly even one to one.*

A less significant argument in support of this "not so empty atom" might be along these lines: Since there is one electron for every proton (both having opposite but equal charge), why then should the electrons *desert* this most favorable position for one which makes no *Biological*

(a)   **(b)**

Fig. 1–13. Restacking The Nucleons

Sense at all, i.e., offspring (electrons) sticking close to "parent" proton—Just as Mother Nature would have it.

Finally, it can be argued that this type of Structure for Atoms may not be so far fetched after all.

First of all, *No One Has Yet Isolated A Free Quark!* And if this model of the Atom's Structure could possibly be correct, it would then have to be that the Quarks are simply *Electrons with Fractional Charges* as long as their orbits are completely *inside* the atom; and *Full Charges* when they are caught at the outer fringes of the atom, i.e., all the electrons of hydrogen atoms and helium atoms, etc., and those electrons orbiting the *outer protons* of all other atoms would be measured as having their *full charge*.

The actual mechanism then, which *determines the density* of all material matter, would simply depend upon the number of *Exposed Outer* protons of each and every atom; wherein, it would actually depend upon the *Repulsive Force* which protons have upon protons from one atom to the next which gives to all matter the extent of its density.

In other words, the material comprising the more massive atoms, (like gold or osmium) have a much lower frequency of *proton to proton repulsive incidents* due to their higher mass number (i.e., giving them larger diameters) and their excess neutrons, thereby keeping them denser, and or *closer together* on average; whereby, as with hydrogen, helium, aluminum, etc., the *repulsion frequency* (due to proton approaches—one to another) is enormously greater since their mass number is so much lower, and there are fewer, or no excess neutrons to help reduce proton to proton approaches. Therefore, the higher the frequency of *proton to proton repulsive encounters,* the less will be the density of the material in question—the best example, of course, being *Hydrogen gas.*

11

What Holds It All Together

What is the force that holds all matter together? Very specifically: what is the *Binding Force* which holds protons and neutrons together in the nucleus?

Let us begin answering this question by citing Newton's first law of motion, which states that: "Every body continues in its state of rest or of uniform motion in a straight line, except insofar as it is compelled by other forces to change that state."

With this in mind, let us take into account the postulate I have proposed in which all matter in the Universe (whether they be electrons, protons, planets, or stars) is eternally traveling through space with a constant rate of acceleration of 2.8 Angstroms per second per second. This being the case, we would then have to believe that it is this very minuscule rate of acceleration, per se, which is the prime factor in the creation of the gravitational force. Would not then this constant rate of acceleration begin to explain for us the reason for Newtons second law of motion being what it is? For it is this constant rate of acceleration *in itself* which violates Newtons First Law, and in fact, creates Newtons Second Law of Motion, F=ma, in which an acceleration is being applied to a mass; and has therefore compelled that mass to change its state of rest or uniform motion. Consequently, *mass* times *acceleration* equals *Force*. And it is this force which precisely creates the *magnitude of g* at the body's surface—and likewise— *the acceleration of free fall* due to gravity, in an "equal quantity of units" in m/s^2.

As for the forces of gravity, it is well known that when an object is raised it will acquire potential energy in an amount determined by: 1-its mass; 2-the height it is lifted to; and 3-the mass of the body it is lifted from. Once released, the object will then possess kinetic energy while descending; and when it contacts the original surface it was lifted from, it will impact with the same amount of energy that had been required to lift it, *its potential energy*.

So then if we lift an object, say a stone, to a certain height, we have performed some work on it. Therefore, since work and energy are equivalent, we have imparted to the object a certain potential energy. The stone retains that potential energy for as long as it is held at that particular height or position. However, it naturally has a tendency to return to its original position of rest. Why? Because it was compelled by *other forces* to change its state of rest and or its uniform motion. Once again however, I would point out that the only reason it *resists a change* is because it and the larger body it appears to be resting upon in the first place (the Earth in this case) are themselves traveling through space, accelerating as they do, at a constant rate of 2.8 Angstroms per second per second. (This in itself represents a force F, on the Earth and Stone in the first place, but that is not the issue being addressed at this time.)

Of course, the key factor in determining the precise amount of potential energy of any raised body from its original position of rest, is in relation to its own mass and the mass of the heavier body in which it is raised from. In other words, a baseball lifted 5 feet from Earth's surface has a greater potential energy than that same baseball lifted from the Moon's surface whose gravitational attraction is only about one-sixth of Earths. On Jupiter, a baseball lifted 5 feet would have more than twice the potential energy of the baseball lifted 5 feet on Earth since Jupiter's gravitational pull at its surface is more than twice the Earth's pull.

So what are we saying? We are saying that because all material bodies in the Universe are traveling through space with a constant rate of acceleration of 2.8 Angstroms per second per second, then each and every body, *as a consequence of this natural rate of acceleration,* possesses an attracting force, F, which is determined by the simple equation, $F = m_r a_u$. In this equation, F represents the magnitude of g in Newton's per kilogram at a body's surface; m_r represents the body's *Mass of One-Square Meter Radius* (A New unit); and a_u, the Universal Constant of Acceleration. We are speaking of course, *Ideally, of Isolated Spherical Bodies of Uniform Density and Radius.*

Let us consider for example, two separate masses, such as say, one neutron and one proton. There is a gravitational field created upon and around each particle for the simple fact that they are both basically accelerating at the rate of 2.8 Angstroms per second per second as they travel through space in one general direction.

First of all however, in order for gravitational forces to exist, but even more so, *to act,* there must exist at least two separate masses. But then, there can really be no mass at all unless something is first in motion due to acceleration. For it is, after all, motion itself which creates Mass and Energy. And just as Einstein's famous $E = mc^2$ correctly teaches, Energy and Mass are the equivalent of one another and are therefore interchangeable.

So now, we have two separate masses for the sole reason that they are both accelerating. Once again then, let us identify these two masses as consisting of one neutron and one proton. Their density is about 10^{20} kg per cubic meter, and they each have a radius of about 1.5×10^{-16} meter (these are my own estimates). We would then have from the equation ($F=m_r a_u$), a force at their surfaces of approximately 10^{-6} N/kg. That is, the magnitude of g at the surface of the proton or neutron is about 10^{-6} Newton per kilogram in the same way that the magnitude of g at the Earth's surface is 9.8 Newtons per kilogram. (See Chapter 13 for additional details.)

12

Speculations

On Quasars, Tides, the 3-K Radiation and the Compass

If all matter throughout the Universe is indeed in a constant state of acceleration, it then becomes likely that there may possibly exist other explanations for a multitude of phenomena with which we are familiar. A few of these will be discussed in this chapter and will briefly be touched upon. In particular, I will be "Highly Speculating" in attempting to show that, because I believe all matter is naturally and eternally accelerating at a constant rate, there may be other, more meaningful explanations for phenomena such as: the celebrated 3-K Background Radiation, Quasars, the Earth's Ocean Tides, the Sun's Surface Activities, and possibly even the Rotational Action in a Compass.

On The 3-K Background Radiation

Can it be that the 3-K Blackbody Radiation is due directly to our own basic velocity through space of 300,000 km per second, rather than the "Left Over Radiation of a so-called Big Bang?"

I refer here to the Cosmic Background Radiation detected by Arno Penzias and Robert Wilson of Bell Labs in 1965. This blackbody radiation is usually referred to as the 3-K Background Radiation, but as is known, it is actually closer to 2.7° Kelvin.

What is significant about this background radiation is that it appears to fill all space and is almost exactly the same temperature in all directions. I use the term "almost" because of some measurements made in the late 1970's by Richard Muller and George Smoot which indicated that this background radiation was not as perfectly smooth as was originally believed. Instead they found that it was about 0.0035°K warmer than average in the direction of the constellation Leo, and at exactly 180° in the opposite direction, toward the constellation Aquarius, it was found to be 0.0035°K cooler than average.

It was also found that this background radiation declined in temperature in a smooth manner as measured from the warm spot in Leo to the cool spot in Aquarius. This *temperature variation* was then properly accounted for as being due to the Earth's overall motion through the Cosmos—that is—we are traveling in the Milky Way from Aquarius toward Leo at a speed of about 390 kilometers per second. Specifically, this speed corresponds to a temperature excess of 0.0035° Kelvin. (Author's Note: these details concerning the cosmic background radiation temperature variation can be found in the text entitled *Universe*, by William J. Kaufman, published 1985, by W. H. Freeman & Co.).

The main point I wish to emphasize is that since this temperature excess of 0.0035°K is specifically equal to a speed of 390 kilometers per second, then the whole of the 2.7° background radiation we detect can very precisely be equated to what I propose to be the Earth's basic speed through space of 299,792 kilometers per second; since on the basis of 0.0035°K being equated with a velocity of 390 kilometers per second, then likewise, the velocity of Earth through space of 299,792 kilometers per second precisely equates with a temperature excess of 2.69 degrees kelvin! Therefore, I would strongly suspect that the total of the 2.7° background radiation may actually be due, "not" to the remnants of a so-called Big Bang, but rather, to the Earth's basic and present velocity of 299,792 kilometers per second through space. For when one calculates these numbers, it will be found that 390 kilometers per second divides into 299,792 kilometers per second 768.7 times. If we then multiply this *"768.7-to-one ratio"* by .0035 degrees, we arrive at a product of 2.69 degrees kelvin.

In other words, the celebrated 2.7° Cosmic Background Radiation may very simply be due to the fact that the Earth is indeed speeding through space with a present velocity of approximately 300,000 kilometers per second, thereby providing us with a more direct, more realistic reason to understand and explain the *real origin* of the Background Radiation.

The satellite COBE, (Cosmic Background Explorer) is soon due to be launched, unless it has already been. One of its functions will be to find and confirm these Deviations in the Background Radiation similar to the existing evidence from the investigations of Smoot & Muller.

On Quasars

Is there a logical explanation for Quasars? Imagine if you will—two hot stars such as our Sun—each rushing toward each other from the outer edges of two different expanding universes. Each star then collides head on with the other at a speed of say 300,000 kilometers per second, possibly even as high as 600,000 km per second, if not more. The energy, and or total light output from two hot stars colliding head on at 300,000 kilometers or more per second each, should in theory be equal to, or similar to, what astrophysicists presently observe when they are looking at Quasars. For if what I have already proposed (as concerns a universal acceleration) are *facts to be reckoned* with, then there may well be good reason to suspect that Quasars are *in fact* two colliding stars somewhere near the outer edges of our own expanding Universe and another foreign universe—both of whose outer edges have finally come to expand into each others turf—something unavoidable in time if more than one universe exists. The energy released from such a collision can probably be calculated close enough to its real value to confirm whether or not it does in fact compare with what we actually now observe with Quasars—*a classical example of the energy release according to* $E=mc^2$!

On Ocean Tides and Solar Surface Activity

Another interesting speculation concerns the Earth's *Ocean Tides*. For example, what relation might exist between Ocean Tides and Earth's 24-hour rotation as it *continuously accelerates* through space in one general direction? That is, what effect would a 2.8 Angstrom per second per second eternal rate of acceleration upon the Earth have on the waters of the Earth's oceans as each point on Earth swings around to the specific direction of it's basic forward thrust through space, that is, in the opposite direction of $t=0$?

Would not the plain fact be that an accelerating body, *containing water,* would simply displace the water from its position of equilibrium? And additionally, if the accelerating body is also rotating, would not the displacement of the water be affected at different locations, *in line,* and *in time,* with the rate of rotation of the accelerating body,

Low Tide. And when a body of water is displaced at one point on the accelerating body by X-amount, again, Low Tide, should not a like displacement take place 180° circumvent of that body of water in the opposite manner, *High Tide?*

Of course, this all brings into question the Moon and its supposed gravitational effect on the Earth. Does the Moon's gravity really play *that substantial part of the tide maker* as such? Or is it really this other side of the coin—the *Acceleration* of all Material Bodies in the Universe at the rate of 2.8 Angstroms per second per second? This would appear to be an interesting area to be investigated.

Also, what relation might exist between Solar Prominences and other Solar Surface Activity as the Sun itself continuously accelerates through space at 2.8 Angstroms per second per second, rotating as it does, one full revolution every 27–31 days?

There are still many unanswered questions concerning, amongst other things, Solar Prominences. But if indeed the Sun does basically accelerate through space at 2.8 Angstroms per second per second, then this constant rotation in itself might supposedly explain the Sun's Prominences, just as constant rotation of the Earth (as suggested above) may also explain the *slight prominences* which we refer to as *ocean tides*.

In any case, where is there a massive body located (in respect to the Sun's position in space) which can actually cause such large solar prominences? Neither the Earth's pull of gravity, nor the rest of the planets in the solar system can have the grossly enormous effect upon the Sun such as we observe of solar prominences. There must then be something else, and a Sun "eternally accelerating at 2.8 Angstroms per second per second" may just possibly provide that impetus!

One factor which could be explained of the Sun (if it is indeed accelerating) would be that which concerns its loss of mass. It is said that the Sun is losing mass at a rate of four million tons per second. However, if it is indeed accelerating, albeit only 2.8 Angstroms per second per second, then according to $E=mc^2$ its mass should constantly be increasing. If this should be the case, then in all probability some of its lost mass (due to its radiation output) should be replaced, at least to some extent, if not all, due to its constant increase in velocity of 2.8×10^{-10}m/s^2; since in Einstein's famous energy equation, $E=mc^2$, energy and mass are interchangeable. And since c^2, as I propose, is nothing more or less than the square of a body's *actual velocity* through space, then it stands to reason that the Sun's mass is constantly increasing as a result of its constant increase in velocity. Of course, as stated previously, it is constantly losing mass due to its radiation output, so that the net effect is that its mass should not actually be decreasing by the large amounts it is presently believed to be.

On The Rotational Mechanism Of A Compass

The final item of interest that I will briefly touch upon is in connection with the *Compass*. This Universal Constant of Acceleration, which I have claimed appears to be the *basic motive force* in the Universe, may also give us some clue to the basic mechanism behind the manner in which a compass responds to change in direction. In other words, what connection might there be between *the rotation of the compass needle*, and north, south, east and west, for a body such as Earth, accelerating through space in one general direction *as it constantly rotates*. Can this not be the sole reason why a compass needle appears to move as the compass is rotated?

These then are just some of the many items for *speculation* which can arise once it is recognized that all matter may be accelerating through space at some finite rate, albeit, a distance per second per second equal only to the length of an average size molecule.

This rate of increase in velocity to an electron however *Is Big Business*. It means, as previously stated, that the electron (whose point-like diameter is estimated to be about 10^{-18} meter) is eternally accelerating at a rate approximately 67-million times *each second* its own length, i.e., it is constantly accelerating at about 10^{-18} meter per 10^{-8} sec per 10^{-8} sec.

In closing this chapter, I present for further speculation the following equation on the Fine Structure Constant utilizing the Universal Constant of Acceleration, a_u, for an elapsed time of one-second:

$$FSC = (a_u/1.0 \text{ Sec})^2 c^2 \tag{15}$$
$$FSC = (2.791 \times 10^{-10})^2 (2.998 \times 10^8)^2$$
$$FSC = 6.999 \times 10^{-3}$$

(UTILIZING THE STANDARD EQUATION)

$$FSC = \mu_o c e^2 / 2h \tag{16}$$
$$FSC = (4\pi \times 10^{-7})(2.998 \times 10^8)(1,602 \times 10^{-19})^2 /$$
$$(2.0 \times 6.626 \times 10^{-34})$$
$$FSC = 7.297 \times 10^{-3}$$

13

GUT's and TOE's

Grand Unified Theories and Theories Of Everything! That is what physicists the world over are in search of—GUT's and TOE's—a Grand Unified Theory in which a *single force* is responsible for creating all the forces of nature; a Theory of Everything which they demand must encompass *through one set of equations* all of the important physical phenomena of nature.

This book has purported to promulgate such a theory by having introduced a handful of equations utilizing the Universal Constant of Acceleration, a_u—for if there is one thing in particular which appears to be clear, it is that *the constituent parts of all atoms are in an eternal state of natural acceleration!*

This then would mean that it is *not actually Light* which possesses the basic motion through space, but in fact, it is the *material bodies themselves* which possess the basic motion, such that, here in our own particular space-time frame, it is we—the Earth and everything upon it—that are rushing off through space at the enormous speed we attribute to light. And that it is for this reason, and this reason alone, that light itself exists as it does, *mimicking precisely* our own real and basic velocity through space.

And if in fact the constituent parts of atoms are naturally accelerating, there should then be *no speculation* why we experience a force of attraction at the Earth's surface as we do. So that Gravitation, per se, turns out to be a direct consequence of the fact that all matter is eternally accelerating at a rate of 2.8 Angstroms per second per second; and that the propagation and precise velocity of light are a direct result of the emitting body's velocity through space.

The fact that all the constituent parts of atoms are in a natural and eternal state of acceleration should also explain why *heavy objects* and *light objects* would fall to the ground (in empty space) at the same rate—it is simply because they fall as *Individual Protons and Neutrons*. So that protons and neutrons do not care how we or mother nature have packaged them—volume for volume, *they each fall to the ground as a Single Entity*—whether packaged as *styrofoam*, or packaged as *lead*. It explains too why the Sun, Earth, Moon, and all other massive bodies accelerate through space at the same rate of 2.8 Angstroms per second per second. It is because they too accelerate as *Individual Protons*. It simply does not matter how they are packaged—*Protons Never Lose Their Identity!*

I have shown the Universal Constant of Acceleration, a_u, to be $2.790\ 625\ 94 \times 10^{-10}$ m/s². However, I realize this does not necessarily mean it is precise. This of course is so, because it is constructed with the use of other constants and values, most notably, the constants c; h; H_o; the mass of the proton; and the densities, radiuses, and magnitudes of g for the planets of the solar system. All were important factors in fine tuning the value for a_u. And so, the degree

GUT's AND TOE's

of their accuracy naturally determined the degree of accuracy for a_u.

In closing, there remains one item which deserves to be elaborated upon. That is the *Strong Nuclear Force*—the force which binds Protons and Neutrons together to give stability to the atom.

As I had attempted to show in a prior chapter, the Strong Force can be envisioned when one realizes that the g Force, that is, the acceleration due to gravity at the "surfaces of protons and neutrons" (although only about 10^{-7} meter per second per second) is still, to each, an Enormous Attractive Force when one considers that at 10^{-7} meter per second per second, this means that the gravitational pull (the magnitude of g at the surface) of one proton upon one neutron is *equivalent* to a rate of fall—which when measured in *proton diameters*—would be *hundreds of millions of times the proton's diameter per second per second*.

Compare that to a Human Body falling to Earth in which the accelerated displacement rate is equal to *only six (6) body lengths per second per second*. So that in this case, the body accelerates each second, only a small minute fraction of the total diameter of the Earth.

If however, we consider two bodies such as one proton and one neutron (instead of a human body and a body the size and mass of Earth), the proton and neutron fall towards each other, so to speak, at a rate equal to *hundreds of millions* of times per second per second their own lengths. In addition, the body they fall to (or are attracted to) is the same size and mass as the falling or attracted body itself.

When we are reminded then that the gravitational force is so weak that at the Quantum Level it has *no significance* at all, nor nothing whatsoever to do with the Strong Nuclear Force, this would appear to be *Wholly Incorrect;* for although the magnitude of g at the surfaces of Protons and Neutrons is very minuscule (10^{-7} Newton) in comparison to gravitational forces we ourselves are used to, those *quantum level minuscule forces* are however, enormously more powerful on a *proton to neutron scale*, than the attractive forces are on a "Large Body Scale" such as with planets and people.

So when all is said and done, it may well turn out to be that the Strong Nuclear Force is really the result of this *tiny amount* of eternal energy—the Universal Constant of Acceleration, a_u,—in which each and every particle throughout the Universe is inherently accelerating; and that it is this *minuscule source of eternal energy* which makes the Universe tick—which makes $E=mc^2$ actually $E=mv^2$; which gives the propagation of light its unique quality of a *built in velocity;* and which causes the Universe to expand as it does.

And if you ever wondered about the energy which keeps the miraculous pump, *your heart,* working and working and working, you might then be able to get *just a glimpse* of how *God* and *Nature* work, for therein, with the Universal Constant of Acceleration (a_u), might lay the very basis of the Energy Source for the *Biological Processes* as we know them to be . . .

Einstein, as I have advocated in this book, may have unwittingly erred about the characteristic nature of Light's velocity, but he was indeed very correct about various other matters of great significance which, without many of his good works, the material in this book would not have been possible to promulgate and propound as such.

So, for whatever the message in this book may be worth, none of it *valid or otherwise* could have come about without the labors and genius of the giants of science such as Newton, Galileo, Einstein, Hubble, and the many others, past and present, who have made meaningful contributions to science.

Appendix I

Equations In This Book Utilizing The Universal Constant of Acceleration

$a_u = H_o/2$ $\qquad$ $d_e = a_u/PAR_e$

$a_u = F/m_r$ $\qquad$ $d_p = a_u/PAR_p$

$F = m_r a_u$ $\qquad$ $PAR_e = a_u/d_e$

$m_r = F/a_u$ $\qquad$ $PAR_p = a_u/d_p$

$v_{MW} = a_u/t_{MW}$ $\qquad$ $m_e = a_u k/h$

$c_{MW} = a_u/t_{MW}$ $\qquad$ $\Delta m_p = m_p(a_u/k)^2/c^2$

$FSC = (a_u/k)^2 c^2$ $\qquad$ $\Delta m_e = m_e(a_u/k)^2/c^2$

$R = c^2/(a_u/k)^2$ $\qquad$ $FSC = (a_u/1.0 \text{ Sec})^2 c^2$

$h = m_e a_u 2(\pi^2) k$ $\qquad$ $a_u = d_p PAR_p$

$h/2\pi = m_e a_u \pi k$

Appendix II

Constants and values as Used In This Book

*a_u	Universal Constant of Acceleration	$2.790\ 625\ 94 \times 10^{-10}$ m/s²
c	Present Velocity of Light	$2.997\ 924\ 58 \times 10^{8}$ m/s $1.862\ 823\ 96 \times 10^{5}$ miles p/s
h	Planck constant	$6.626\ 075\ 5 \times 10^{-34}$ Joule-Sec
h/2π	Planck constant	$1.054\ 572\ 66 \times 10^{-34}$ Joule-Sec
h/e	Planck constant	$4.135\ 669\ 2 \times 10^{-18}$ keV-Sec
*m_e	Electron Mass	$1.202\ 887\ 49 \times 10^{-25}$ kg
*m_e	Electron Mass	$6.747\ 701\ 05 \times 10^{7}$ keV
m_p	Proton Mass	$1.672\ 623\ 1 \times 10^{-27}$ kg
m_p	Proton Mass	$9.382\ 723\ 1 \times 10^{5}$ keV
*de	Electron Diameter	$4.135\ 669\ 2 \times 10^{-18}$ meter
*dp	Proton Diameter	$2.974\ 217\ 42 \times 10^{-16}$ meter
*G	Gravitational Const	$6.662\ 128\ 69 \times 10^{-11}$ Nm²/kg²

r_E	Earth Radius	$6.371\ 315 \times 10^6$ meters
*P_E	Earth Density	$5.515\ 564\ 55 \times 10^3$ kg/m^3
*m_E	Earth Mass	$5.975\ 383\ 89 \times 10^{24}$ kg
g_E	Earth Surface Gravity	$9.806\ 65 \times 10^0$ m/s^2
F_E	Earth Magnitude of g at its surface	$9.806\ 65 \times 10^0$ N/kg
π		$3.141\ 592\ 65 \times 10^0$
e	Elementary Charge	$1.602\ 177\ 33 \times 10^{-19}$ Coulomb

Hubble Constant, H_o

*57.449 Km Per Sec Per Megaparsec
*17.614 Km Per Sec Per Million Light Years
*10.945 Miles Per Sec Per Million Light Years
* 5.582 Angstroms Per Second Per Light Second

Hubble Constant ($H_o/2$), Adjusted to Account For The Universal Constant of Acceleration, a_u

*28.724 Km Per Sec Per Megaparsec
* 8.807 Km Per Sec Per Million Light Years
* 5.472 Miles Per Sec Per Million Light Years
* 2.791 Angstroms Per Sec Per Light Second

Velocity of Light (present)

$2.997\ 924\ 58 \times 10^8$	Meters Per Second
$2.997\ 924\ 58 \times 10^{18}$	Angstrons Per Second
$1.862\ 823\ 96 \times 10^5$	Miles Per Second

Length of One Light Year (present)

$9.460\ 528\ 404\ 87 \times 10^{12}$ Kilometers
$5.878\ 499\ 792\ 96 \times 10^{12}$ Miles

Megaparsec $= 3.261 \times 10^6$ Light Years
Seconds Per Year $= 3.155\ 692\ 597\ 47 \times 10^7$

*Age of Milky Way Galaxy = 34,042,730,365 years
*Distance of Milky Way Galaxy from (t=0) = 17,021,365,182 LY'S

* My own constants and values

Appendix III

Symbol

a_u	Universal Constant of Acceleration
H_o	Hubble Constant
$H_o/2$	Hubble Constant for a "Changing c"
m_r	Mass of One-Square-Meter Radius (new unit)
v_{MW}	Velocity of Milky Way Through Space (present)
c_{MW}	Milky Way's Velocity of Light (present)
F	Magnitude of g at a body's surface
FSC	Fine Structure Constant
R	Ratio
PAR	Particle Acceleration Rate
G	Newtons Gravitational Constant
g	acceleration due to gravity
h	Planck Constant
m_e	mass of electron
m_p	mass of proton
d_e	diameter of electron
d_p	diameter of proton

Δm_e	mass increase of electron
Δm_p	mass increase of proton
k	Constant of Proportionality
t=0	Time-Zero
π	pi
P	density of a body
r	radius of a body
d	diameter of a body
N	Newton
kg	kilogram
m	meter
m	mass
E	energy
c	Velocity of Light
t	time
s	second
EMF	Electromotive Force
EMR	Electromagnetic Radiation

Appendix IV

Mass-Energy Conversion Table

h	=	$7.372\ 503\ 236\ 8 \times 10^{-51}$	kg
ev	=	$1.602\ 177\ 33 \times 10^{-19}$	Joule
keV	=	$1.602\ 177\ 33 \times 10^{-16}$	Watt-Sec
keV	=	$1.602\ 177\ 33 \times 10^{-16}$	Joule
keV	=	$1.602\ 177\ 33 \times 10^{-19}$	KWSec
keV	=	$4.449\ 763\ 314 \times 10^{-23}$	KWH
keV	=	$1.782\ 662\ 696 \times 10^{-33}$	kg
Joule	=	1.0×10^{-3}	KW Sec
Joule	=	$2.777\ 322\ 605\ 33 \times 10^{-7}$	KWH
Joule	=	$6.241\ 506\ 363 \times 10^{15}$	keV
Joule	=	$1.112\ 650\ 056\ 03 \times 10^{-17}$	kg
KWSec	=	$6.241\ 506\ 363 \times 10^{18}$	kev
KWSec	=	1.0×10^{3}	Joule
KWSec	=	$1.112\ 650\ 056\ 03 \times 10^{-14}$	Kg

2.8 ANGSTROMS

KWH	=	3.60059×10^{6}	Joule; Watts
KWH	=	$2.247\ 310\ 539\ 66 \times 10^{22}$	keV
KWH	=	$4.006\ 196\ 665\ 4 \times 10^{-11}$	kg
kg	=	$1.356\ 391\ 401\ 7 \times 10^{50}$	h
kg	=	$5.609\ 568\ 166\ 94 \times 10^{32}$	keV
kg	=	$8.987\ 551\ 787\ 36 \times 10^{16}$	Joule
kg	=	$8.987\ 551\ 787\ 36 \times 10^{13}$	kw Sec
kg	=	$2.496\ 133\ 074\ 59 \times 10^{10}$	kw Sec

Bibliography

Adler, Irving. *Inside The Nucleus.* Signet Books, New York, 1964.

Asimov, Isaac. *The Collapsing Universe.* Pocket Books, New York, 1977.

Asimov, Isaac. *The Neutrino.* Avon Books, New York, 1975.

Asimov, Isaac. *The Subatomic Monster.* Mentor Books, New York, 1985.

Barnett, Lincoln. *The Universe and Dr. Einstein.* William Morrow Co., New York, 1966.

Barrow, John D. & Silk, Joseph. *The Left Hand of Creation.* Basic Books, Inc., New York, 1983.

Bergamini, David. *The Universe.* Time-Life Books, New York, 1969.

Bernstein, Jeremy. *Einstein.* Penguin Books, New York, 1985.

Bernstein, Jeremy. *Three Degrees Above Zero.* Mentor Books, New York, 1986.

Berry, Michael. *Principles of Cosmology and Gravitation.* Cambridge University Press, New York, 1978.

Boslough, John. *Stephen Hawking's Universe.* Quill/William Morrow, New York, 1985.

Calder, Nigel. *Einstein's Universe.* Penguin Books, New York, 1980.

Calder, Nigel. *The Key To The Universe*. Penguin Books, New York, 1977.

Capra, Fritjof. *The Tao of Physics*. Bantam Books, New York, 1984.

Capra, Fritjof. *The Turning Point*. Bantam Books, New York, 1988.

Charon, Jean. *Cosmology, Theories of the Universe*. McGraw Hill, New York, 1975.

Clark, Ronald W. *Einstein, The Life and Times*. Avon Books, New York, 1972.

Cole, K. C. *Sympathetic Vibrations*. Bantam Books, New York, 1985.

Coleman, James A. *Modern Theories of the Universe*. Signet Books, New York, 1963.

Cooper, Necia G. and West, Geoffrey B. *Particle Physics*. Cambridge University Press, New York, 1988.

Crease, Robert P. and Mann, Charles C. *The Second Creation*. Collier Books/MacMillen, New York, 1983.

D'Abro, A. *The Rise of the New Physics*. Dover Publications, New York, 1952.

Davies, P. C. W. *Space and Time in the Modern Universe*. Cambridge University Press, New York, 1977.

Ebbighausen, E. G. *Astronomy*. Charles E. Merrill Books, Columbus, Ohio, 1966.

Einstein, Albert. *Essays In Physics*. Philosophical Library, New York, 1950.

Einstein, Albert. *Letters To Solovine*. Philosophical Library, New York, 1986.

Einstein, Albert. *Relativity*. Crown Publishers, New York, 1952.

Einstein, Albert. *Sidelights On Relativity*. Dover Publications, New York, 1983.

Ferris, Timothy. *Galaxies*. Stewart Tabori & Chang Pub., New York, 1982.

Ferris, Timothy. *The Red Limit*. Bantam Books, New York, 1977.

Field, George B.; Arp, Halton; Balecall, John N. *The Redshift Controversy*. W. A. Benjamin, Inc., Reading, Mass., 1976.

Gamow, George. *Gravity*. Anchor Books, Doubleday & Co., New York, 1962.

Gamow, George. *One Two Three—Infinity*. Bantam Books, New York, 1967.

BIBLIOGRAPHY 121

Gamow, George. *Thirty Years That Shook Physics.* Dover, New York, 1966.

Gardner, Martin. *The Relativity Explosion.* Vintage Books, New York, 1976.

Gibbins, Peter. *Particles and Paradoxes.* Cambridge University Press, New York, 1987.

Ginzburg, V. L. *Key Problems of Physics and Astrophysics.* Mir Publishers, Moscow, 1978.

Glashow, Sheldon L. *Interactions.* Warner Books, New York, 1988.

Gleick, James. *Chaos.* Penguin Books, New York, 1987.

Gray, Reginald Irvin. *Unified Physics.* Naval Surface Warfare Center, Dahlgren, VA, 1988.

Gribbin, John and Rees, Martin. *Cosmic Coincidences.* Bantam Books, New York, 1989.

Gribbin, John. *In Search of The Double Helix.* Bantam Books, New York, 1987.

Gribbin, John. *In Search of The Big Bang.* Bantam Books, New York, 1986.

Gribbin, John. *In Search of Schrodinger's Cat.* Bantam Books, New York, 1984.

Gribbin, John. *The Omega Point.* Bantam Books, New York, 1988.

Hey, Tony and Walters, Patrick. *The Quantum Universe.* Cambridge University Press, New York, 1987.

Hoffmann, Banesh. *The Strange Story of The Quantum.* Dover Publications, New York, 1959.

Hoyle, Fred. *Ten Faces of the Universe.* W. H. Freeman and Co., San Francisco, 1977.

Jean, Sir James. *The Growth of Physical Sciences.* Fawcett Publications, Greenwich, CT, 1967.

John, Laurie. *Cosmology Now.* Taplinger Publishing Co., New York, 1976.

Kaku, Michio and Trainer, Jennifer. *Beyond Einstein.* Bantam Books, New York, 1987.

Kaufmann, William J. III. *Relativity and Cosmology.* Harper and Row, New York, 1977.

Kaufmann, William J. III. *Universe.* W. H. Freeman and Co., New York, 1985.

Landau, L. D. and Kitaigorodsky, A. *Physical Bodies*. Mir Publishers, Moscow, 1980.

Landau, L. and Rumer, Yu. *What is the Theory of Relativity*. Mir Publishers, Moscow, 1974.

Levitt, I. M. *Beyond the Known Universe*. The Viking Press, New York, 1974.

Logunov, A. A. *Gravitation and Elemental Particle Physics*. Mir Publishers, Moscow, 1983.

Lorentz, H. A.; Einstein, Albert; Minkowski, H.; Wayl, H. *The Principle of Relativity*. Dover Publications, New York, 1952.

Lusanna, L. *New Trends In Particle Theory*. World Scientific Pub., Singapore, 1985.

McAleer, Neil. *The Mind Boggling Universe*. Doubleday and Co., New York, 1987.

Menzel, Donald H.; Whipple, Fred; L., Vaucouleurs, Gerard de. *Survey of the Universe*. Prentice-Hall, Englewood Cliff, N.J., 1970.

Morris, Richard. *The Nature of Reality*. The Noonday Press, New York, 1987.

Pagels, Heinz R. *Perfect Symmetry*. Bantam Books, New York, 1986.

Pagels, Heinz R. *The Cosmic Code*. Bantam Books, New York, 1984.

Pais, Abraham. *Inward Bound*. Oxford University Press, New York, 1986.

Pais, Abraham. *Subtle is the Lord*. Oxford University Press, New York, 1982.

Parker, Barry. *Einstein's Dream*. Plenum Press, New York, 1987.

Parker, Barry. *Search for a Supertheory*. Plenum Press, New York, 1987.

Reid, R. W. *The Spectroscope*. Signet, New York, 1966.

Riordan, Michael. *The Hunting of the Quark*. Simon and Schuster, New York, 1987.

Rowan-Robinson, Michael. *The Cosmological Distance Ladder*. W. H. Freeman and Co., New York, 1985.

Russell, Bertrand. *The ABC of Relativity*. Signet Books, New York, 1958.

Sagan, Carl. *Cosmos*. Random House, New York, 1980.

Schatzman, E. L. *The Structure of the Universe*. McGraw-Hill, New York, 1976.

Segre, Emilio. *From X-Ray to Quarks*. W. H. Freeman and Co., New York, 1980.

BIBLIOGRAPHY

Singh, Jagjit. *Great Ideas and Theories of Modern Cosmology.* Dover Publications, New York, 1970.

Smorodinsky, Ya A. *Particles, Quanta, Waves.* Mir Publications, Moscow, 1976.

Tilley, Donald E. and Thumm, Walter. *College Physics.* Cummings Publishing Co., Menlo Park, CA, 1971.

Trefil, James S. *From Atoms to Quarks.* Scrubners, New York, 1980.

Trefil, James. *The Dark Side of the Universe.* Anchor Books/Doubleday, New York, 1988.

Watkins, Peter. *Story of the W and Z.* Cambridge University Press, New York, 1986.

Weinberg, Steven. *The First Three Minutes.* Basic Books, New York, 1977.

Whipple, Fred L. *Earth, Moon, and Planets.* Harvard University Press, Cambridge, 1971.

Wilczek, Frank and Devine, Betsy. *Longing for the Harmonies.* W. W. Norton & Co., New York, 1988.

Will, Clifford M. *Was Einstein Right?* Basic Books, New York, 1986.

Zee, A. *Fearful Symmetry.* Collier Books, Macmillan Pub. Co., New York, 1986.

Zukav, Gary. *The Dancing Wu Li Masters.* Bantam Books, New York, 1979.

About the Author

Kenneth Salem has been working since 1968 to convince "Fluid Dynamacists" that Turbulence in Fluid Flows is the result of "Ordinary Acoustical Pressures." In the process of his experimental work he discovered that (contrary to popular belief), individual sub-micron size particles, one-tenth micron and smaller, can be detected with the naked eye. Upon acquiring a U.S Patent for an apparatus he invented to bring into play the particle phenomenon he discovered, he had the honor of displaying the "Particle Detector" to the public at the first ever National Inventors Day Exposition held by the U.S. Patent Office in Washington D.C. Other participants at the show included such prestigious companies as AT&T, Polaroid Corp., and Mazda Corp. with their Wankel Engine.

Salem lives with his wife, Jean, in the historical flood renowned city of Johnstown, Pennsylvannia.

SAN DIEGO